Laboratory Atlas of Anatomy and Physiology

Laboratory Atlas of Anatomy and Physiology

Fourth Edition

Douglas J. Eder, PhD

Southern Illinois University Edwardsville

Shari Lewis Kaminsky, DPM

John W. Bertram

Revision Contributions by:

John R. Waters, MS

Department of Biology
Penn State University

Bruce D. Wingerd, MS

Department of Biology
San Diego State University

Boston Burr Ridge, IL Dubuque, IA Madison, WI New York San Francisco St. Louis
Bangkok Bogotá Caracas Kuala Lumpur Lisbon London Madrid Mexico City
Milan Montreal New Delhi Santiago Seoul Singapore Sydney Taipei Toronto

LABORATORY ATLAS OF ANATOMY AND PHYSIOLOGY, FOURTH EDITION

Published by McGraw-Hill, a business unit of The McGraw-Hill Companies, Inc., 1221 Avenue of the Americas, New York, NY 10020.

This book is printed on recycled, acid-free paper containing 10% postconsumer waste.

2 3 4 5 6 7 8 9 0 QPD/QPD 0 9 8 7 6 5 4

ISBN 0-07-243810-X

Publisher: *Martin J. Lange*
Sponsoring editor: *Michelle Watnick*
Senior developmental editor: *Patricia Hesse*
Director of development: *Kristine Tibbetts*
Marketing manager: *James F. Connely*
Project manager: *Joyce Watters*
Lead production supervisor: *Sandy Ludovissy*
Senior media technology producer: *Barbara R. Block*
Coordinator of freelance design: *Rick D. Noel*
Cover designer: *Rick D. Noel*
Cover image provided by the authors
Senior photo research coordinator: *John C. Leland*
Compositor: *Carlisle Communications, Ltd.*
Typeface: *10/12 Goudy*
Printer: *Quebecor World Dubuque, IA*

The credits section for this book begins on page 169 and is considered an extension of the copyright page.

Library of Congress Cataloging-in-Publication Data

Laboratory atlas of anatomy and physiology / Douglas J. Eder, Shari Lewis Kaminsky, John W. Bertram, John Waters.—4th ed.
p. cm.
Previous ed. cataloged under: Eder, Douglas J.
Includes index.
ISBN 0-07-243810-X (hbk. : alk. paper)
1. Human anatomy—Atlases. 2. Human physiology—Atlases. I. Eder, Douglas J.

QM25.L33 2004
611—dc21

2002155909
CIP

www.mhhe.com

CONTENTS

DEDICATION

To Suzanne, Nicholas, Daniel, and Mary Ellen

ACKNOWLEDGMENTS

We obtained all animal material pictured from Nasco Company, Ft. Atkinson, Wisconsin. Cat cadavers were skinned at the factory and packed in a non-formaldehyde preservative. At our request, Nasco personnel selected particularly well-injected cadavers for us; we thank them for this service. We would also like to thank our colleague, Dave Bolt of West Hills College, who reviewed our animal disections.

We would like to acknowledge the valuable contributions of the reviewers for the fourth edition who provided detailed recommendations for improving the manual.

Shawn Bjerke
Fergus Falls Community College

John Capeheart
University of Houston—Downtown

Danielle Desroches
William Paterson University of New Jersey

Dalia Giedrimiene
Saint Joseph College

Shelley A. Kirkpatrick
Saint Francis University

William E. Montgomery
College of Southern Maryland

David P. Sogn Mork
St. Cloud State University

Devonna Sue Morra
Saint Francis University

Janet E. Steele
University of Nebraska at Kearney

CHAPTER 1

Histology

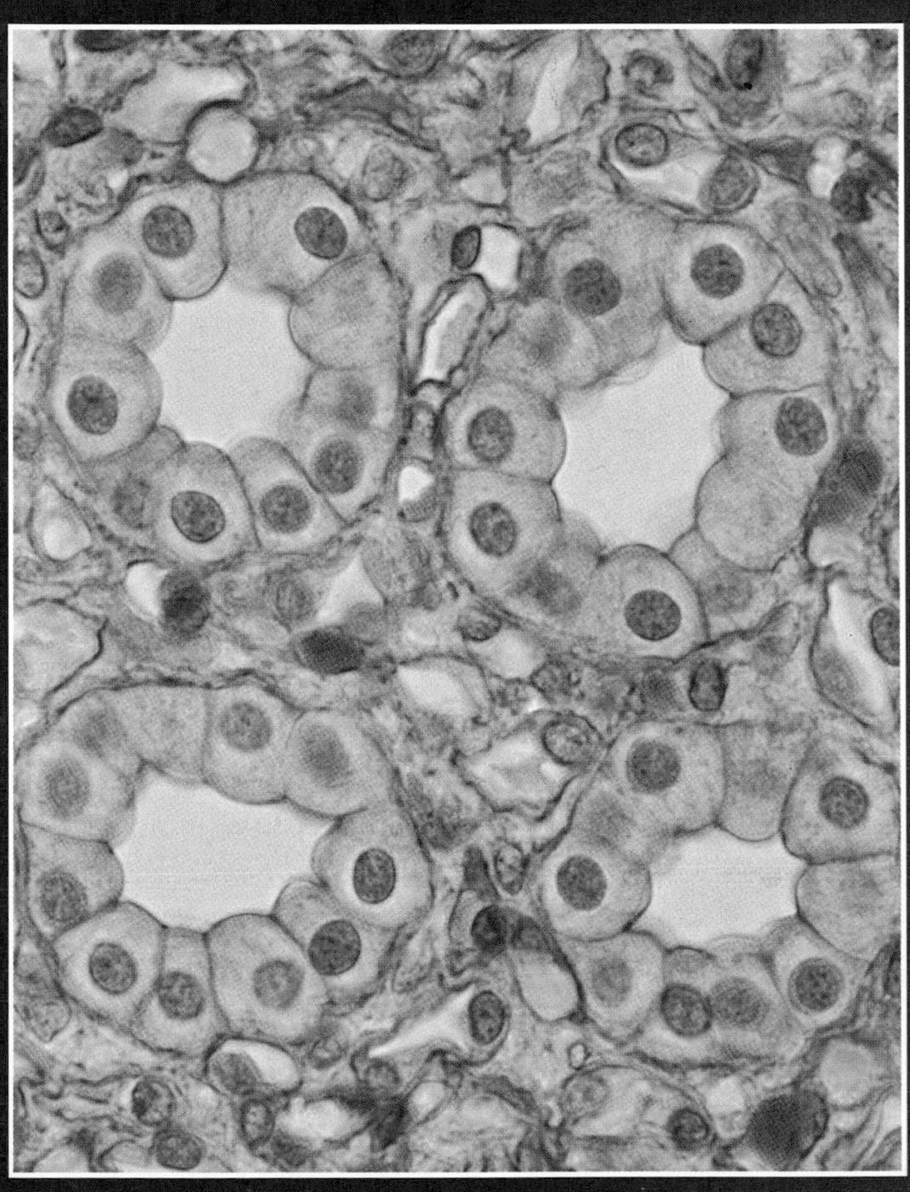

Simple Cuboidal Epithelium

Figure 1-1
Interphase Nuclear envelope intact with chromatin visible. (×250)

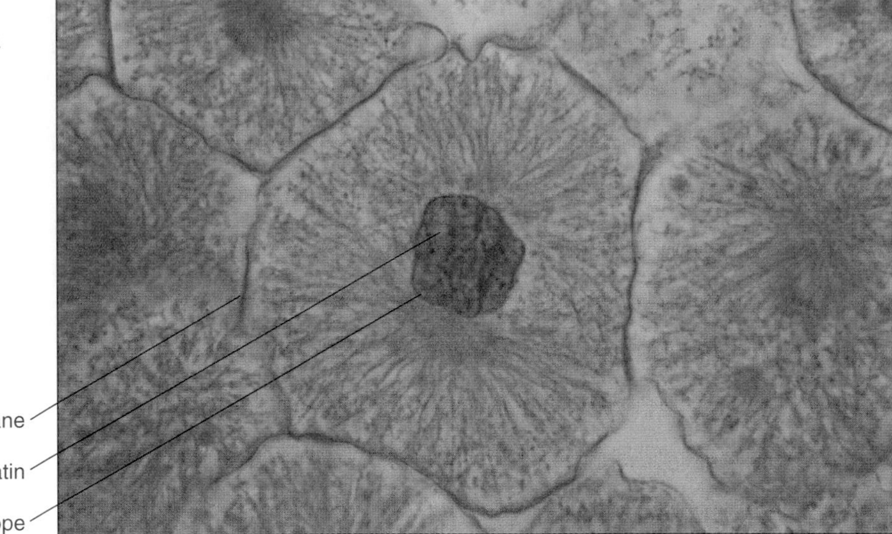

Figure 1-2
Prophase Duplicated chromosomes condensed into visible strands; nuclear envelope absent. (×250)

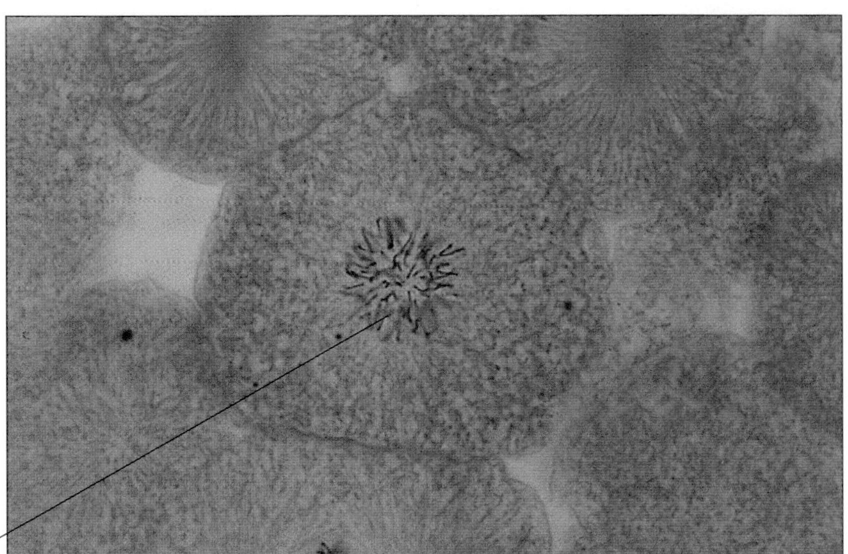

Figure 1-3
Metaphase Darkly stained chromosomes positioned by microtubular framework to align at cell equator. Spindle fibers and astral rays visible. (×250)

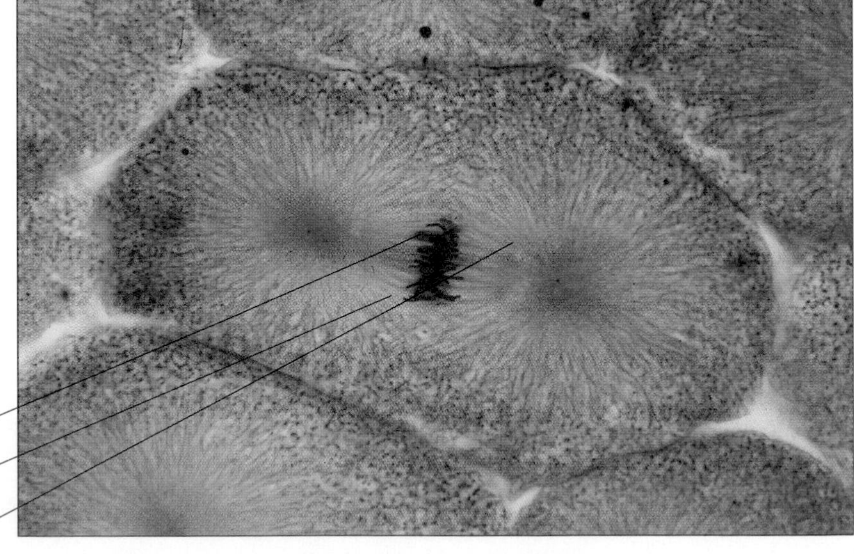

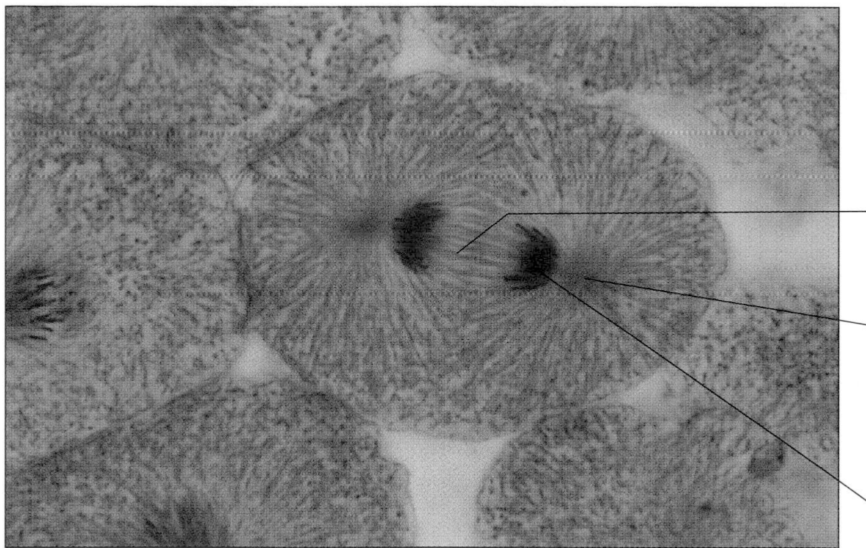

Figure 1-4
Anaphase Darkly stained chromosomes move to opposite poles under microtubular influence. Spindle fibers and astral rays visible. (×250)

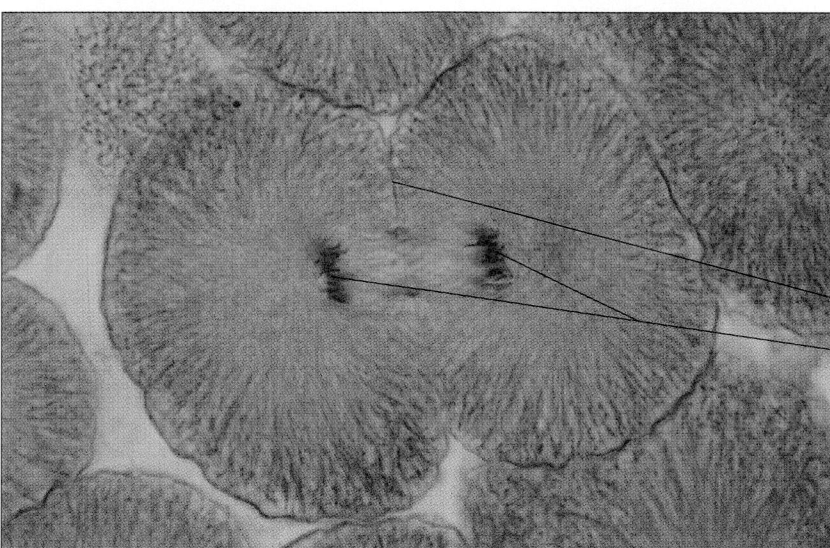

Figure 1-5
Telophase Separated chromosomes lose microtubular attachments. Belt of actinomyocin forms at equator, assists in formation of new cell membranes and cytokinesis. Cleavage furrow forms two daughter cells. (×250)

Figure 1-6
Simple Squamous Epithelium Single layer of flat cells covering a surface. From human omentum. (×250)

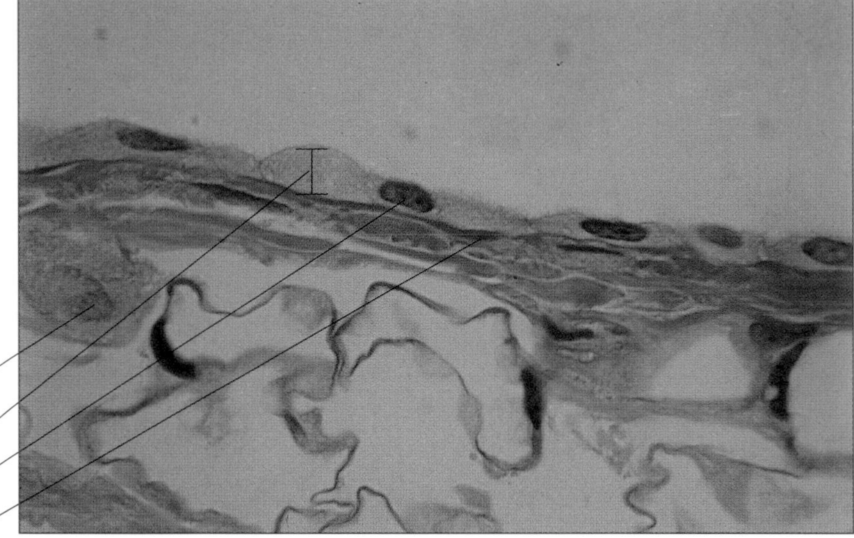

Figure 1-7
Simple Squamous Epithelium Surface view of flattened cells. Human mesothelium. (×250)

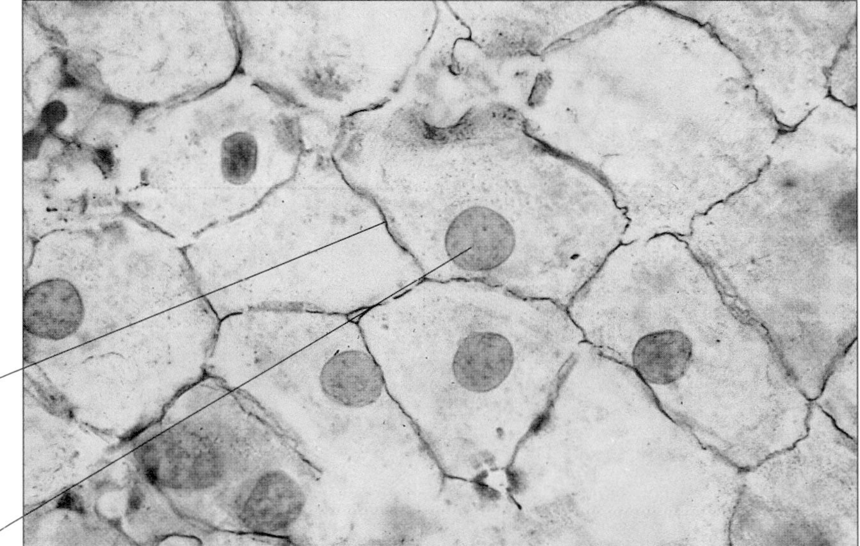

Figure 1-8
Simple Cuboidal Epithelium Although not strictly cube shaped, cuboidal cells are roughly equidimensional in length, width, and depth. Single layer of cells lining surface of kidney tubules. Cross section. (×250)

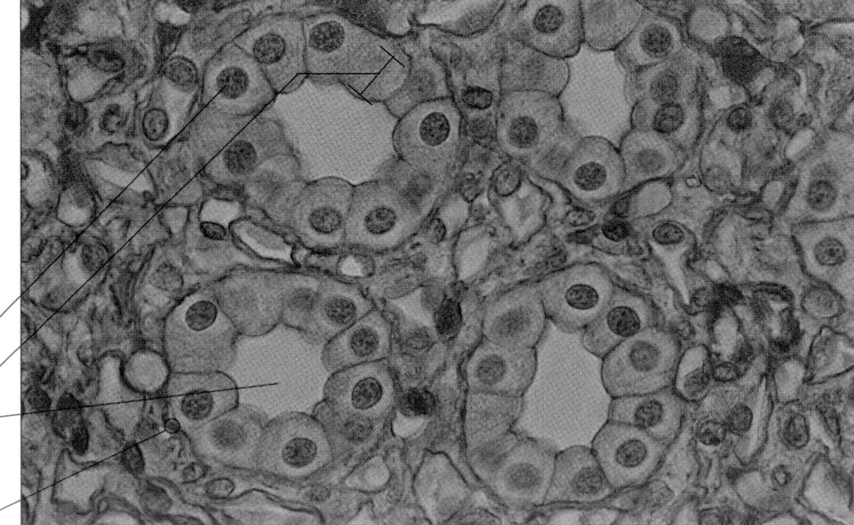

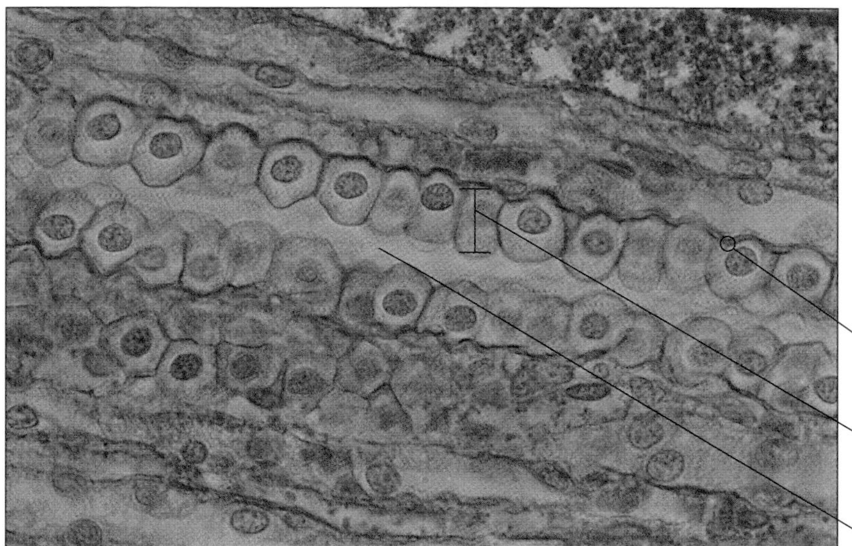

Figure 1-9
Simple Cuboidal Epithelium Longitudinal section of kidney tubule. (×250)

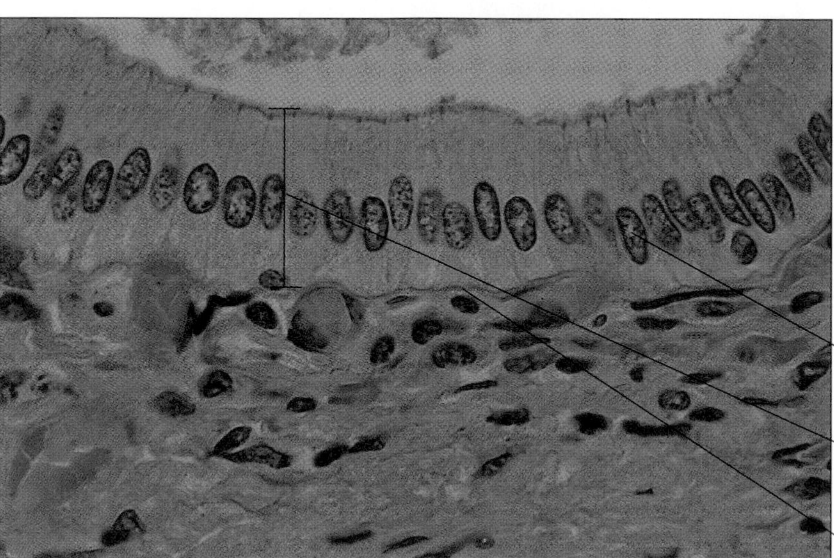

Figure 1-10
Simple Columnar Epithelium Cellular height is much greater than width or length. Nuclei generally appear in a row. From pancreatic duct. (×250)

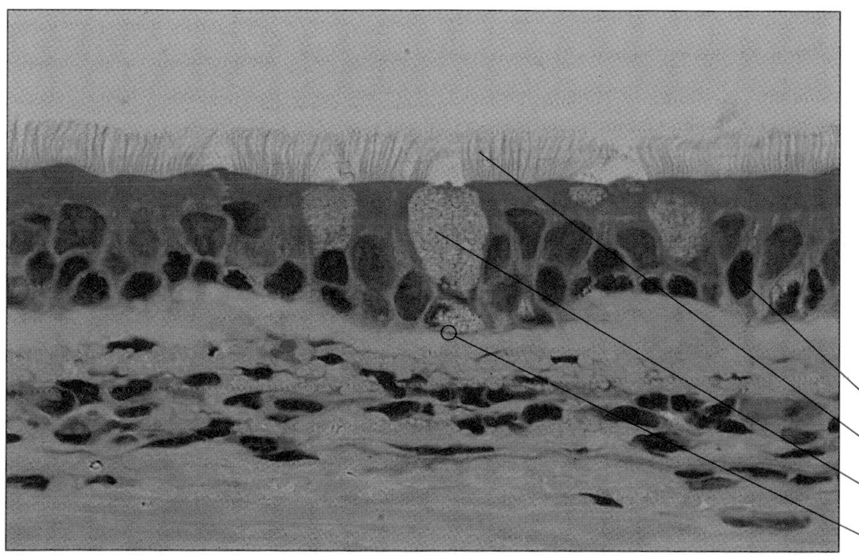

Figure 1-11
Pseudostratified Ciliated Columnar Epithelium Nuclei appear to lie in two rows, but in fact all cells in single layer are in contact with basement membrane. Section shows well-defined cilia, three goblet cells, basement membrane, underlying connective tissue. From monkey trachea. (×100)

Figure 1-12
Pseudostratified Ciliated Columnar Epithelium Section shows cilia, multiple layers of nuclei, basement membrane, underlying connective tissue. From human trachea. (×250)

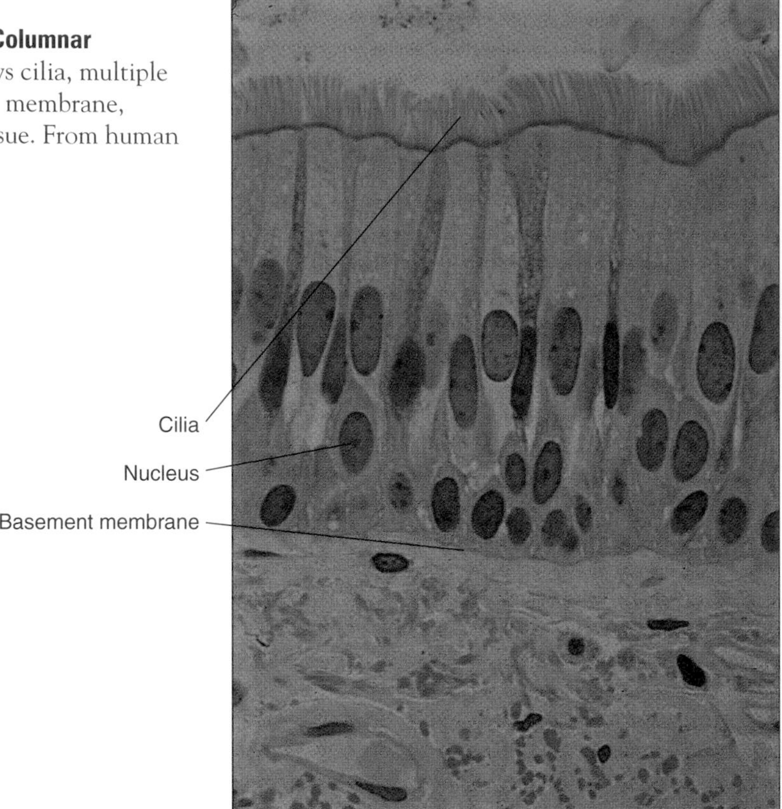

Figure 1-13
Stratified Squamous Epithelium Flattened cells at surface change to less flattened morphology in deeper layers. Oral cavity of rabbit. (×100)

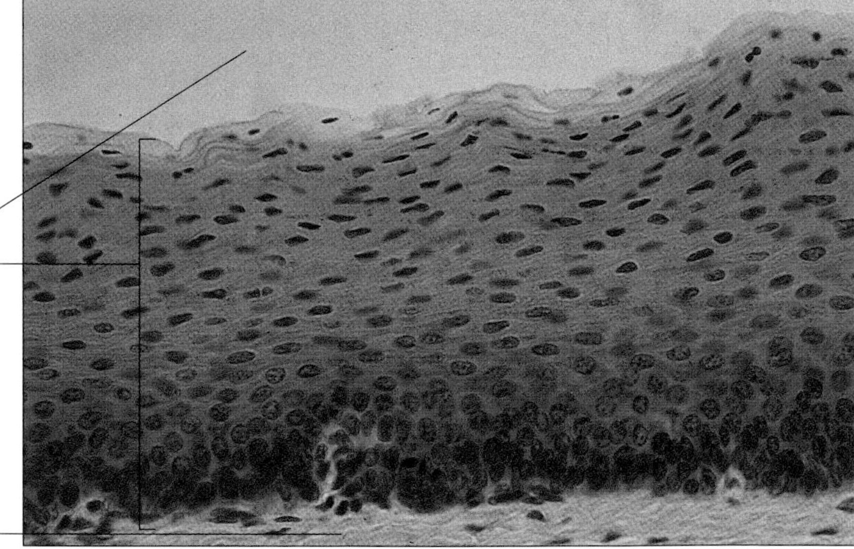

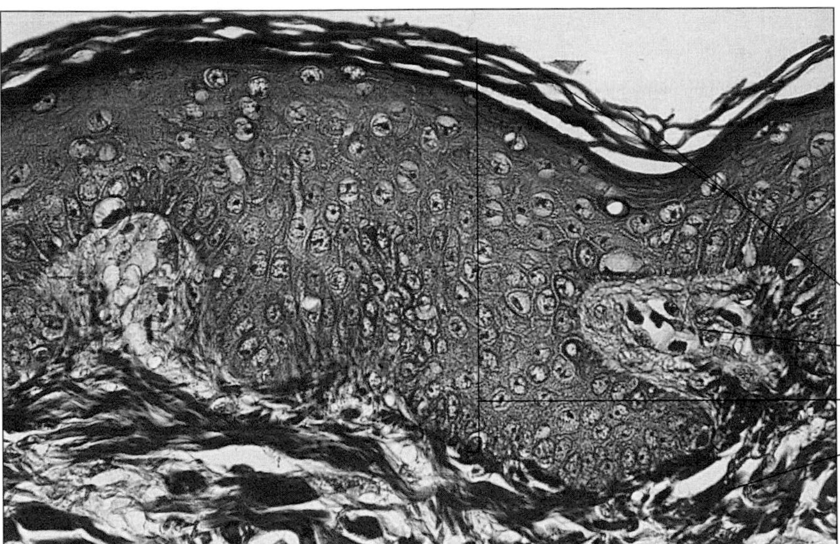

Figure 1-14
Stratified Squamous Epithelium Flattened, keratinized cells at surface show variations in form in deeper layers. From human skin. (×100)

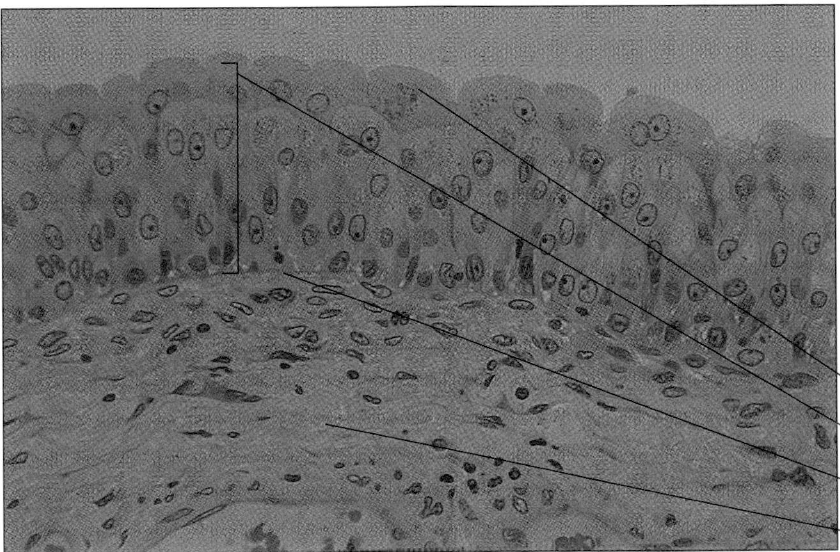

Figure 1-15
Transitional Epithelium from Urinary Bladder Umbrella cells stretch and flatten as bladder fills. Basement membrane separates epithelium from underlying connective tissue containing blood vessels. (×250)

Figure 1-16
Elastic Connective Tissue Extracellular elastic fibers running parallel in a plane. Structure permits tissue elasticity and recoil. From aorta. (×100)

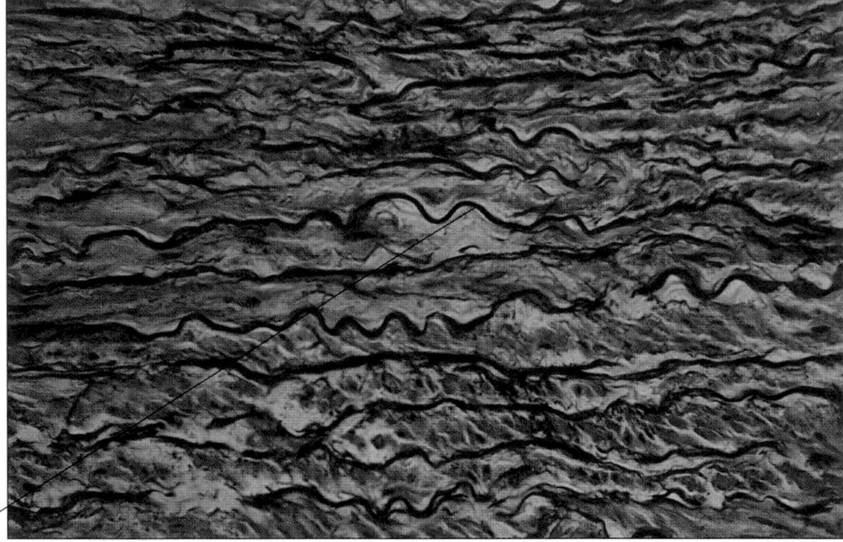

Figure 1-17
Reticular Connective Tissue Mesh of reticular fibers appears as dark lines; provides scaffold for cellular organization of this lymph node. (×250)

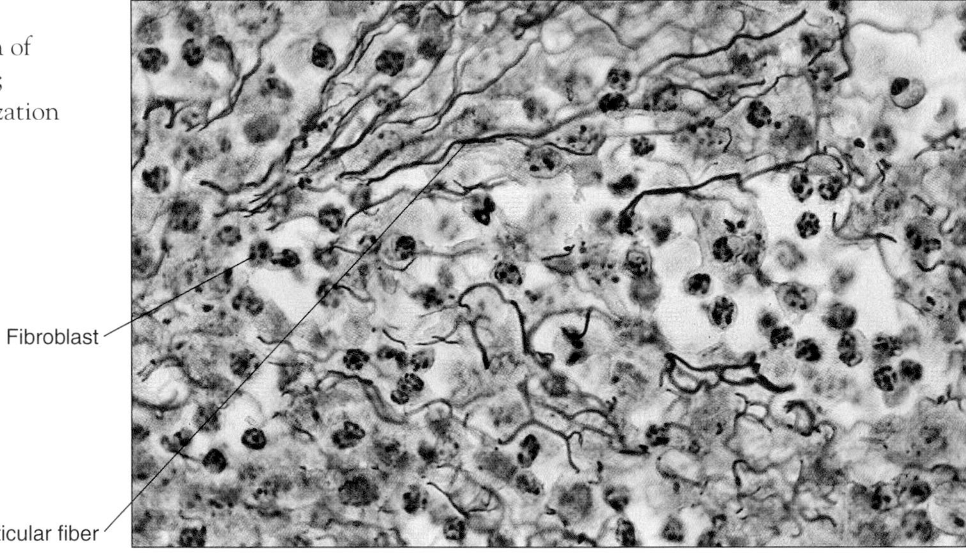

Figure 1-18
Loose (Areolar) Connective Tissue Pink bands of collagen fibers run in all directions through intercellular spaces of subcutaneous tissue, which permit flexible resistance to mechanical stress. (×100)

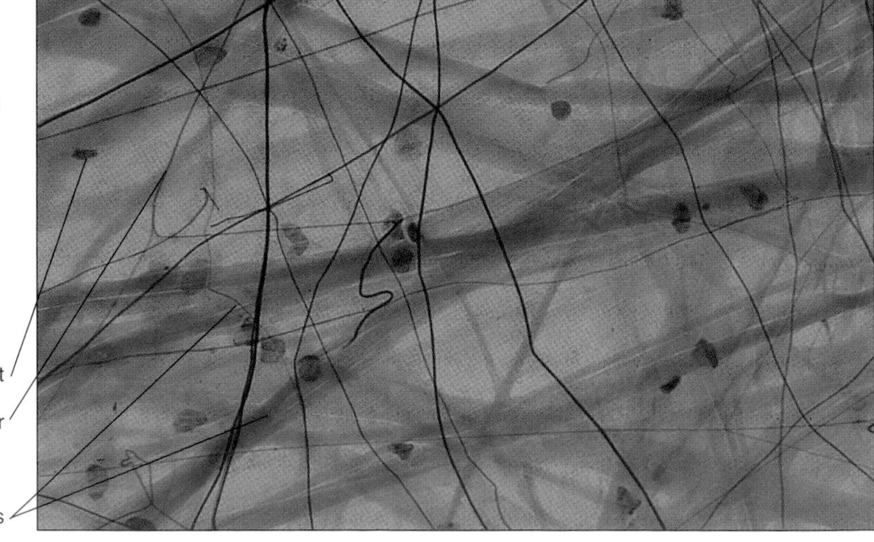

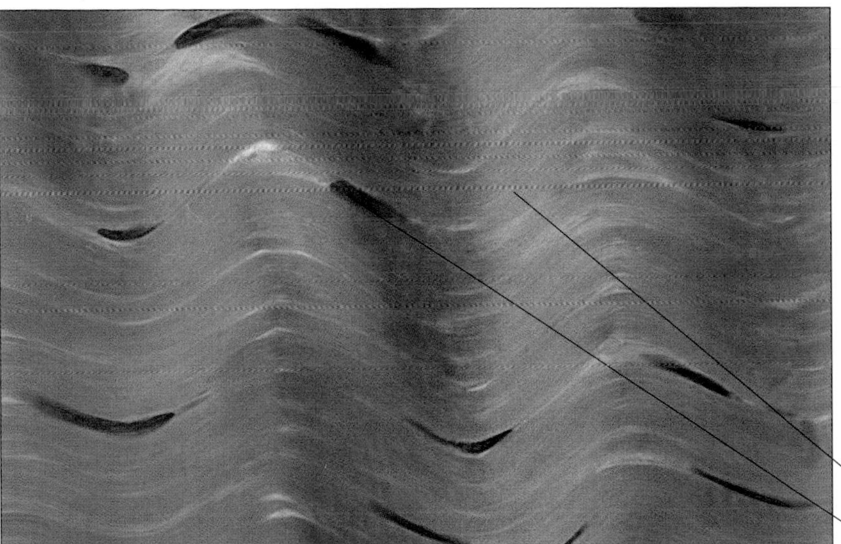

Figure 1-19
Dense Regular Connective Tissue Bands of collagen fibers running in regular, parallel rows resist mechanical stress mainly along course of fibers. Monkey tendon. (×250)

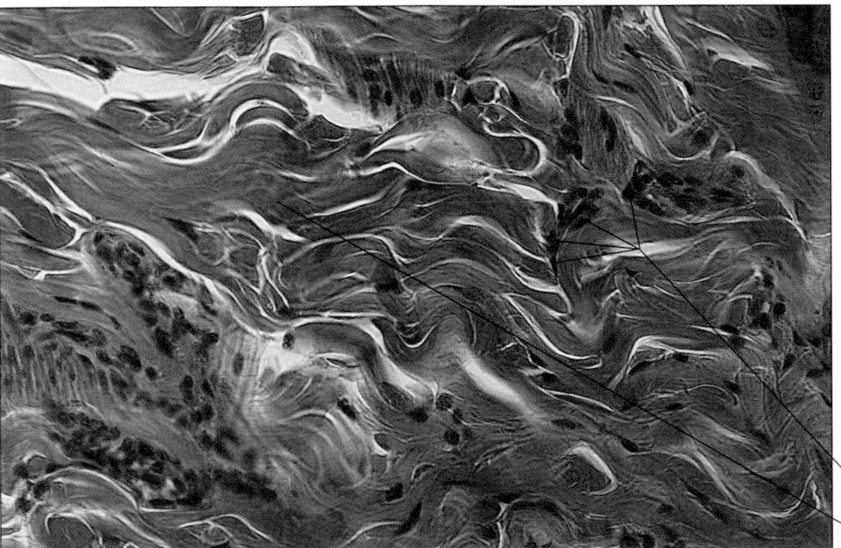

Figure 1-20
Dense Irregular Connective Tissue Bands of collagen running in irregular rows give multidirectional tensile strength. Collagen-secreting fibroblasts appear throughout. (×100)

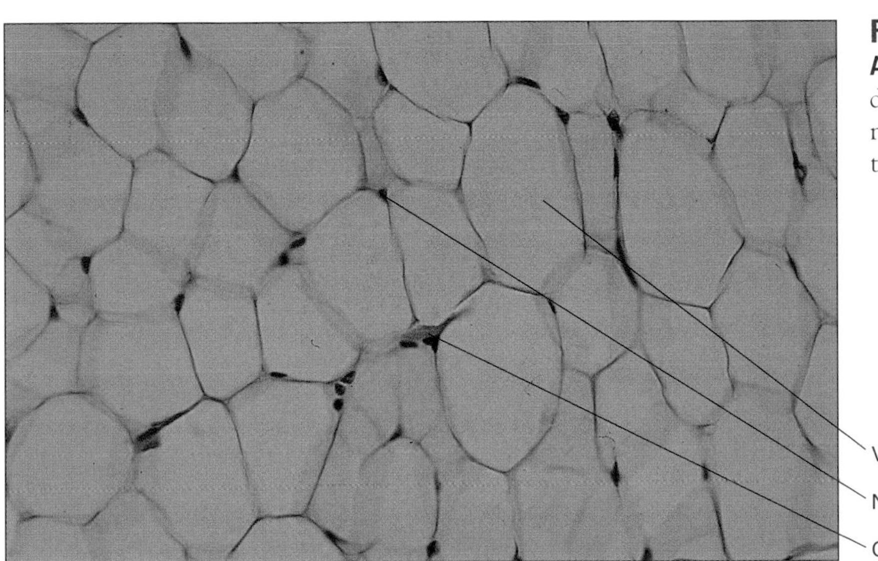

Figure 1-21
Adipose Tissue Large, polyhedral vacuoles dominate small, eccentrically located cell nuclei of adipocytes. Fine capillaries run through tissue. (×100)

Figure 1-22
Fibrocartilage Cell nests of chondrocytes in territorial matrix surrounded by coarse extracellular fibers. (×250)

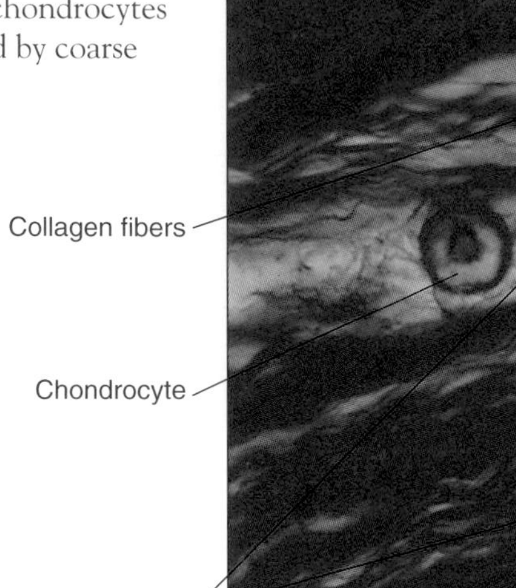

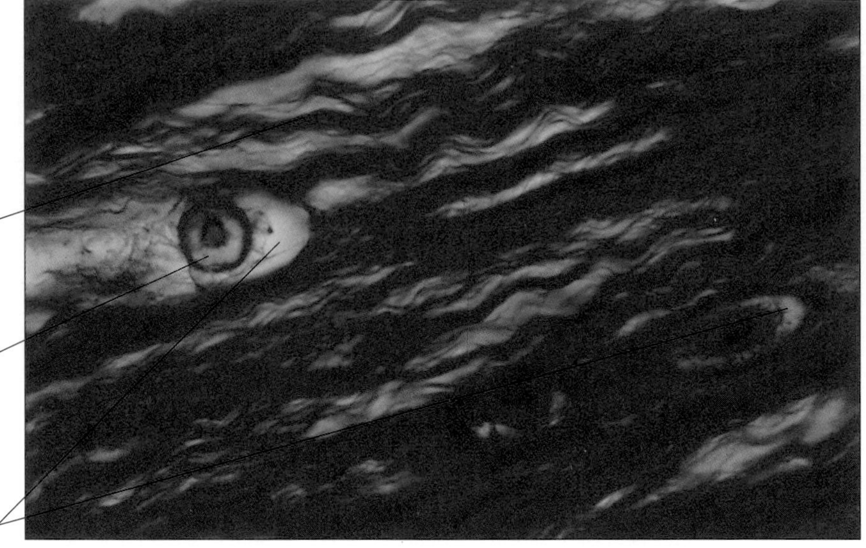

Figure 1-23
Hyaline Cartilage During interstitial growth, cartilage cells often form small clusters and move apart as they secrete extracellular matrix. (×100)

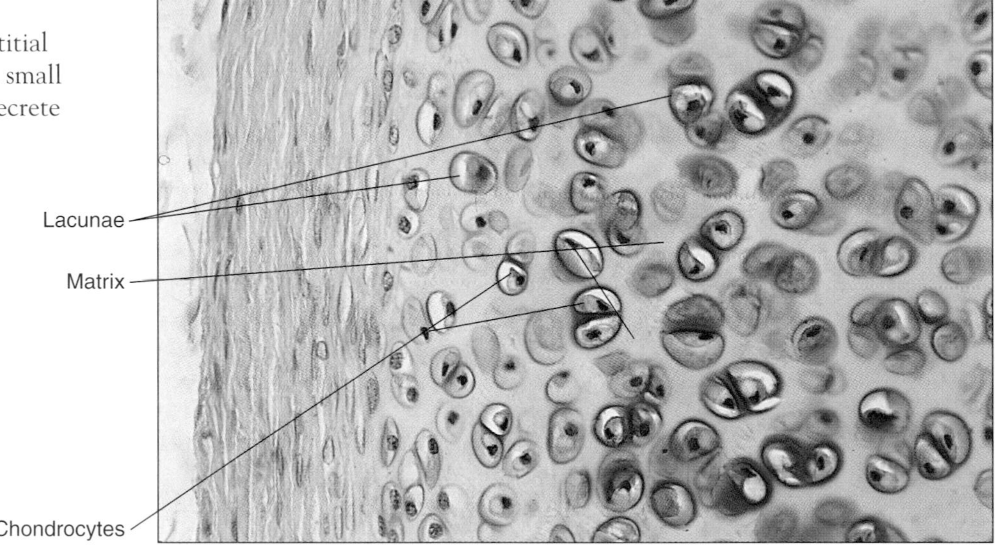

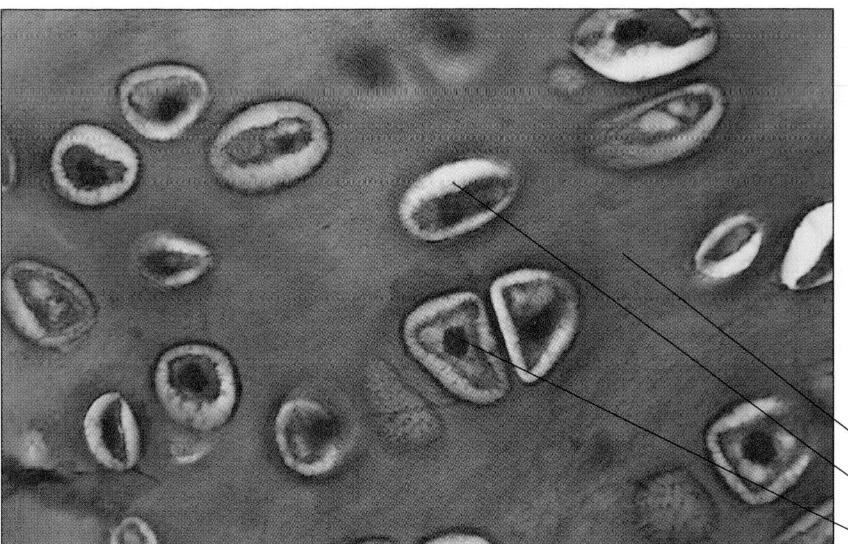

Figure 1-24
Hyaline Cartilage Artifactual vacuolation forms characteristic lacunae around chondrocyte cell bodies. From trachea. (×250)

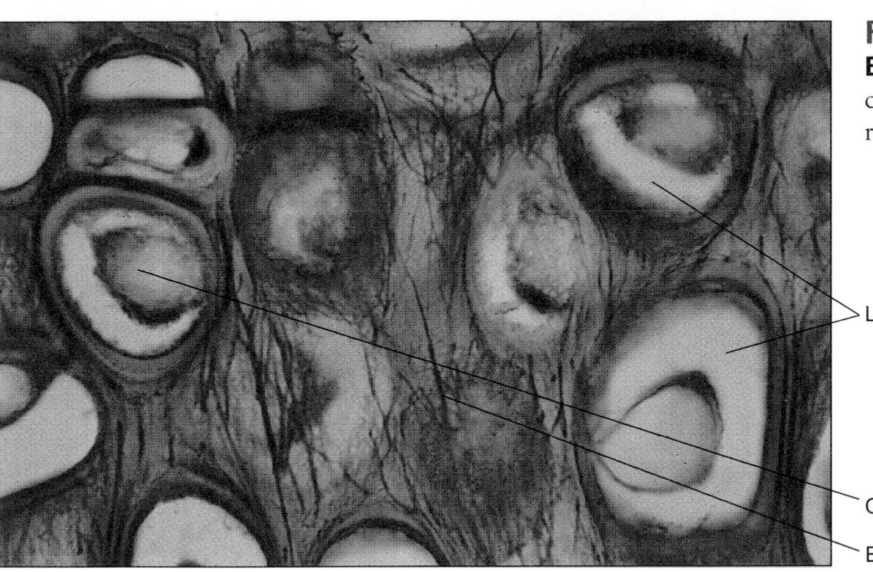

Figure 1-25
Elastic Cartilage Extracellular matrix contains elastic fibers that confer elastic recoil to this tissue. (×250)

Figure 1-26
Skin Thick, keratinized, multilayered **stratum corneum** rests atop grainy **stratum granulosum** (stratum lucidum not clearly evident). **Stratum spinosum,** composed of irregularly shaped cells with indistinct nuclei, lies atop single, clearly nucleated **stratum basale.** Human palm. (×100)

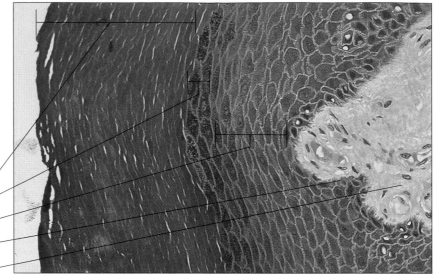

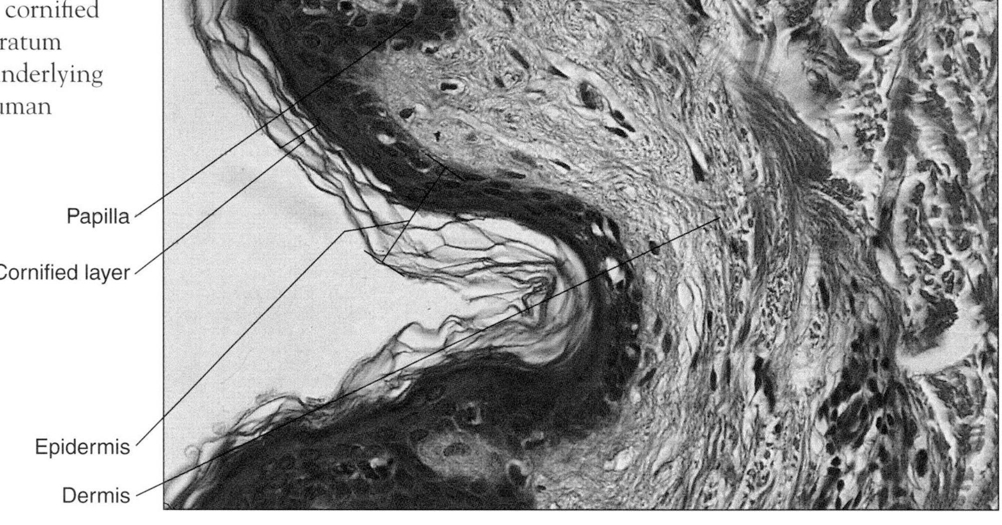

Figure 1-27
Skin Squamous epidermis with cornified layers overlying darkly stained stratum basale and connective tissue of underlying dermis. Single papilla visible. Human scalp. (×100)

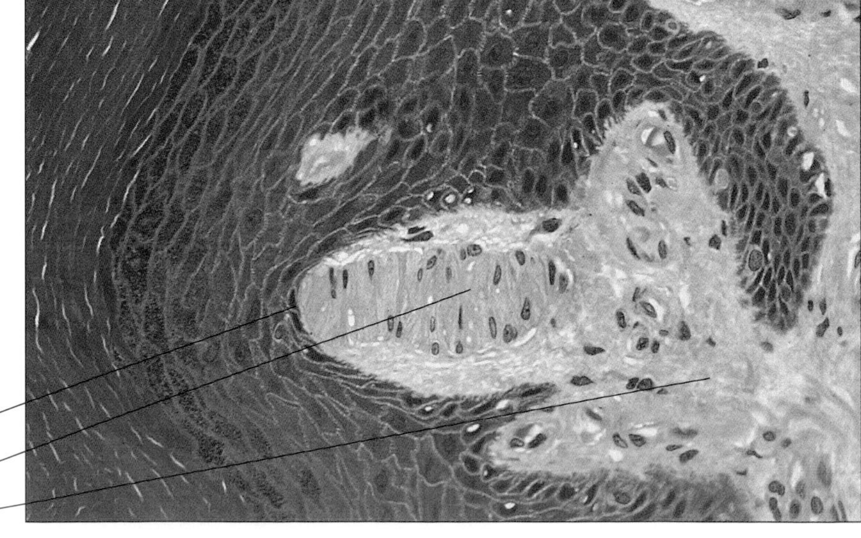

Figure 1-28
Meissner's Corpuscle in Dermis Elongated oval body located in dermis just below stratum basale is thought to be responsible for part of fine touch reception. (×100)

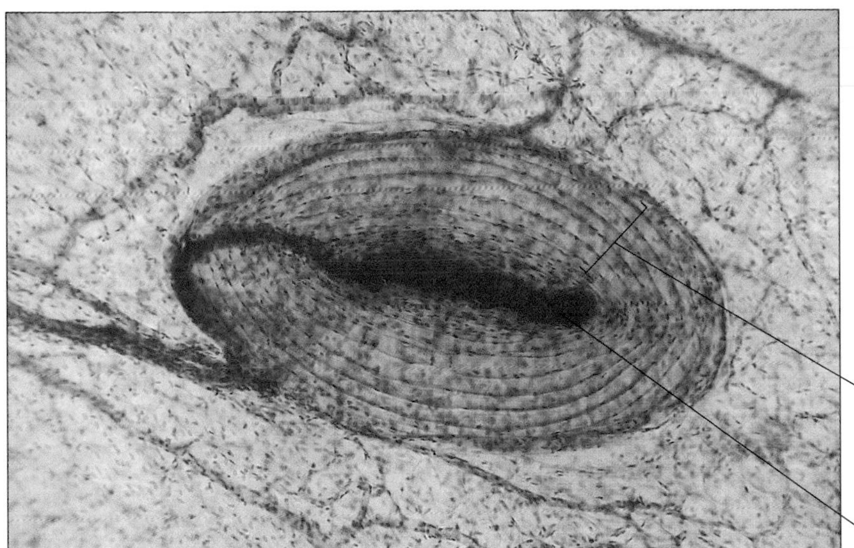

Figure 1-29
Pacinian Corpuscle Encapsulated nerve ending found deep in dermis and throughout interior of body detects pressure. (×25)

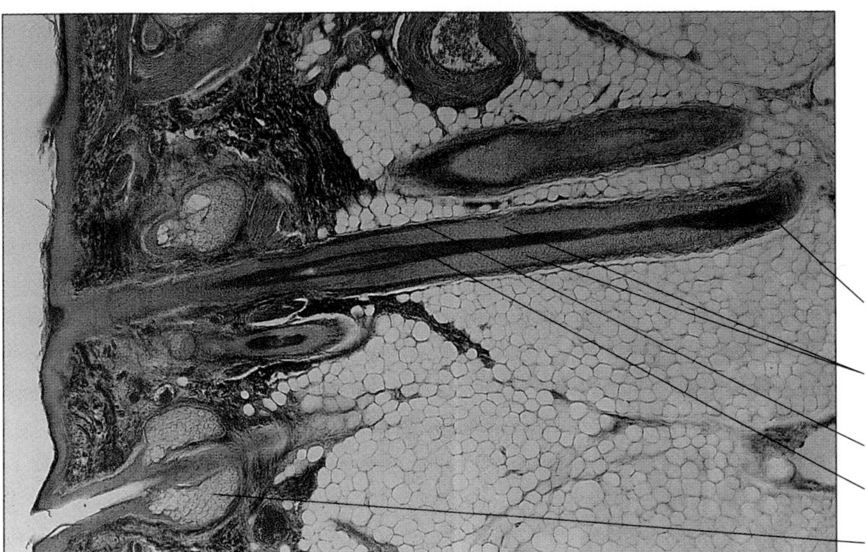

Figure 1-30
Human Scalp with Hair Follicle Follicle root, with sheath embedded in pale adipose tissue, has sebaceous glands surrounding it near surface. (×10)

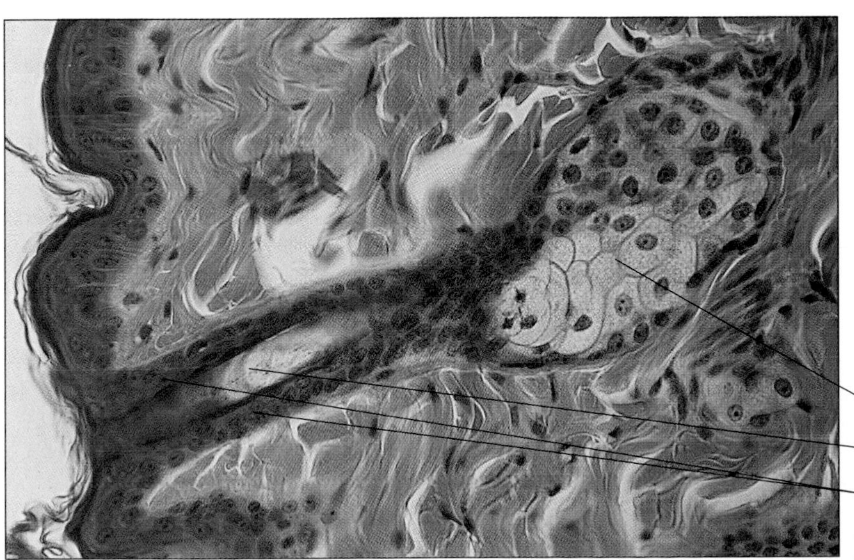

Figure 1-31
Detail of Sebaceous Gland Nucleated germinative cells at base of gland mature and accumulate lipid. At duct, they degenerate and lyse to release their oily product, sebum. (×100)

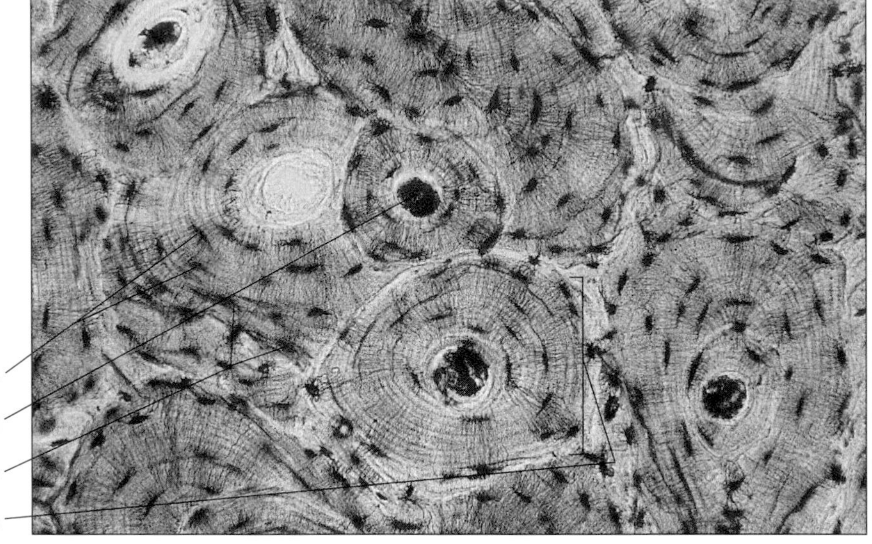

Figure 1-32
Compact Bone Center of "tree ring" structure, osteonic Haversian canal contains blood vessel. Osteocytes imprisoned in small, dark lacunae surrounding central osteonic canal receive nutrition and communicate via canaliculi, or little canals. Human. (×50)

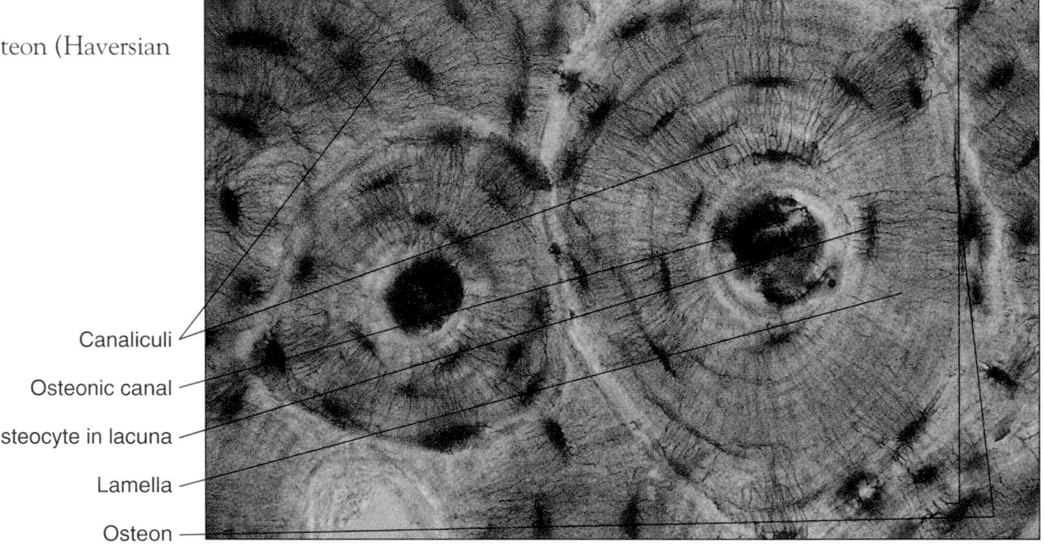

Figure 1-33
Detail of Compact Bone Osteon (Haversian system) evident. (×100)

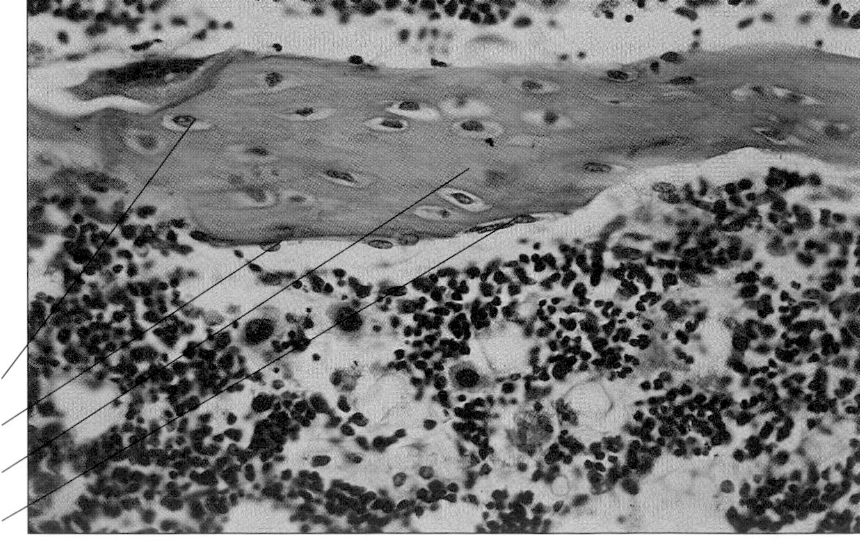

Figure 1-34
Cancellous (Spongy) Bone Osteoblasts on spongy bone are engaged in secretion of new bony matrix. (×100)

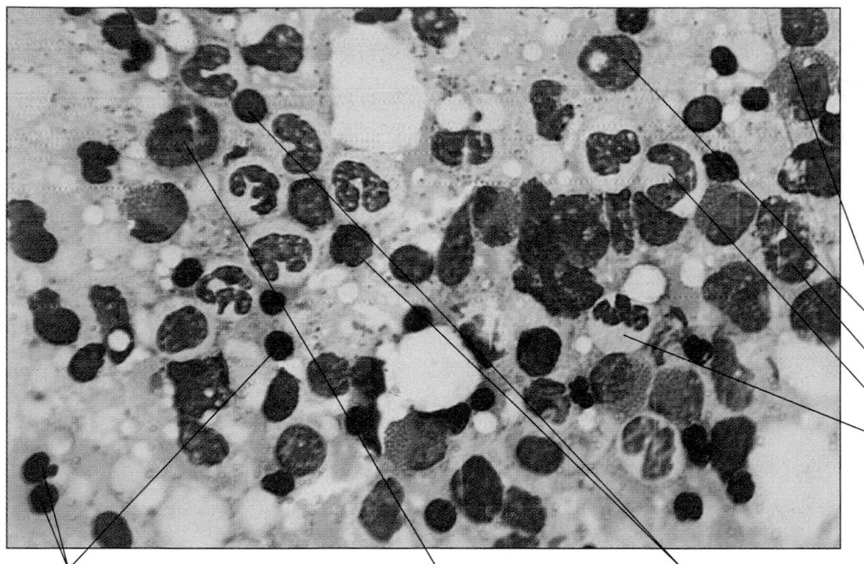

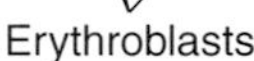

Figure 1-35
Red Bone Marrow Medullary cavity in the head of long bones of the adult contains stem cells, precursors to red blood cells, and white blood cells and platelets. Human. (×250)

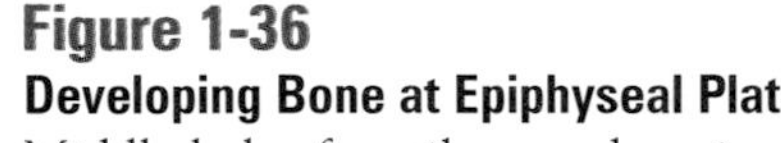

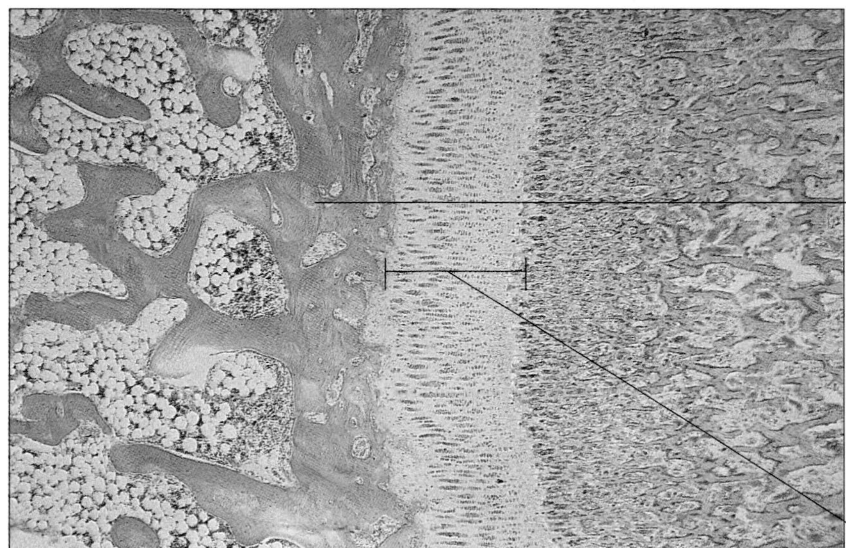

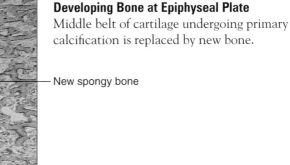

Figure 1-36
Developing Bone at Epiphyseal Plate Middle belt of cartilage undergoing primary calcification is replaced by new bone.

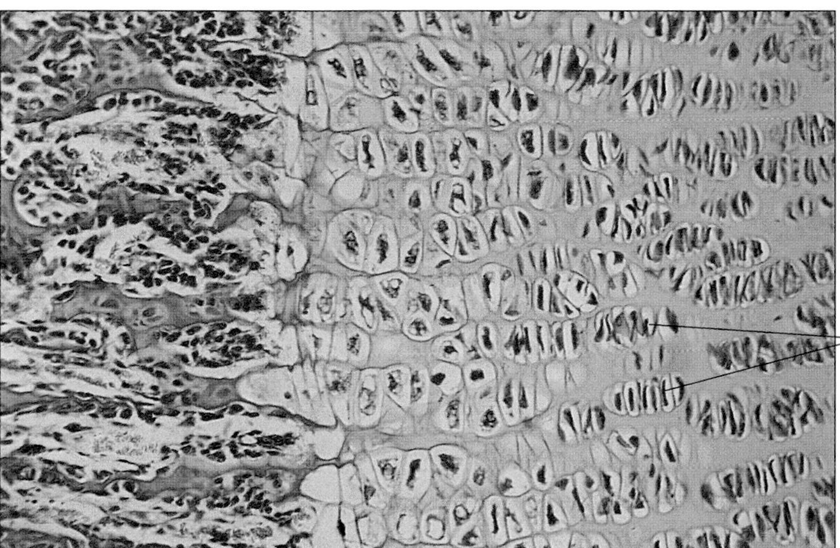

Figure 1-37
Detail of Epiphyseal Plate Epiphyseal plate cartilage at right transforms into zones of proliferating chondrocytes with primary ossification occurring on their calcified remnants. Newly formed bone appears at left. (×50)

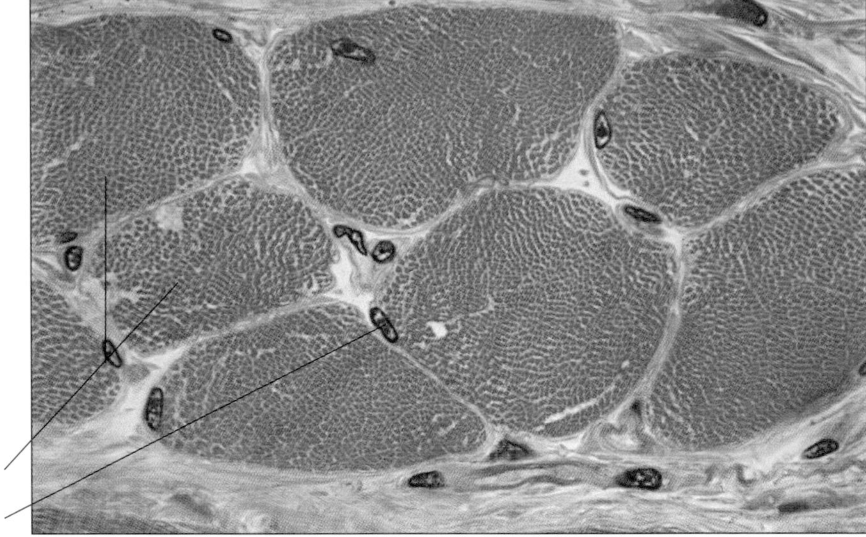

Figure 1-38
Striated (Skeletal) Muscle (Cross Section) Eccentrically located multiple nuclei accompany individual cells (fibers), each of which contains many myofibrils. Human tongue. (×250)

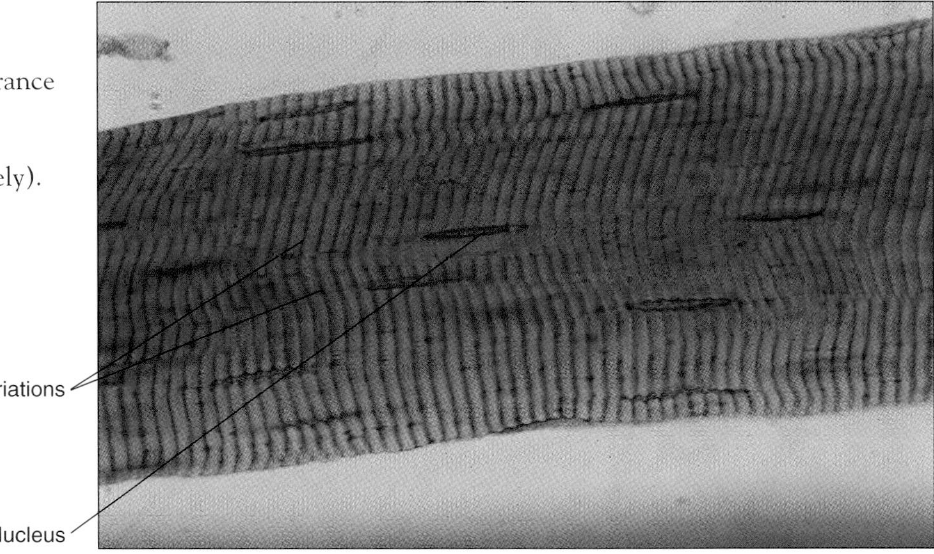

Figure 1-39
Striated (Skeletal) Muscle Fiber (Longitudinal Section) Banded appearance arises from regular arrangement of overlapping bundles of thick and thin filaments (myosin and actin, respectively). Eccentrically located nuclei are thin and elongated. (×250)

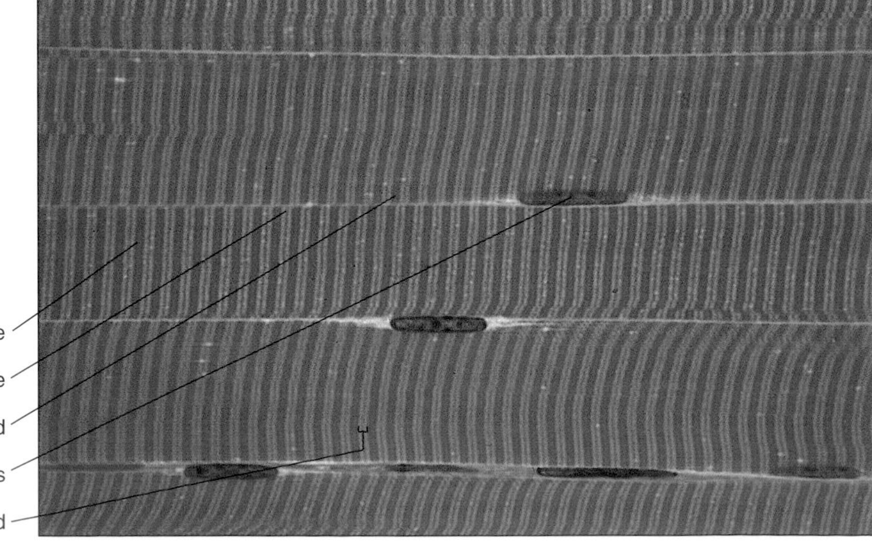

Figure 1-40
Striated (Skeletal) Muscle Fibers (Longitudinal Section) Each light *(I)* band has a dark *(Z)* line through it. Each dark *(A)* band has a light *(H)* zone through it. (×250)

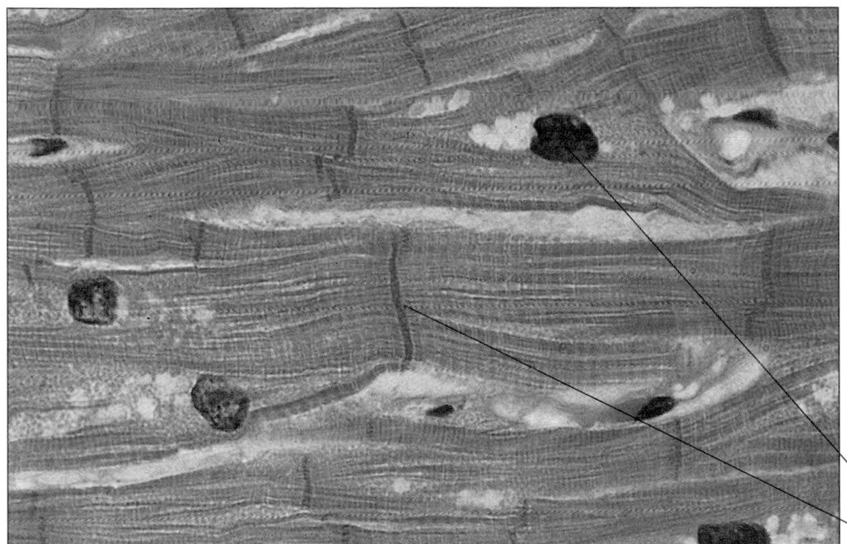

Figure 1-41
Cardiac Muscle (Longitudinal Section)
Striated muscle fibers branch and anastomose at junctions marked by dark intercalated disks. (×250)

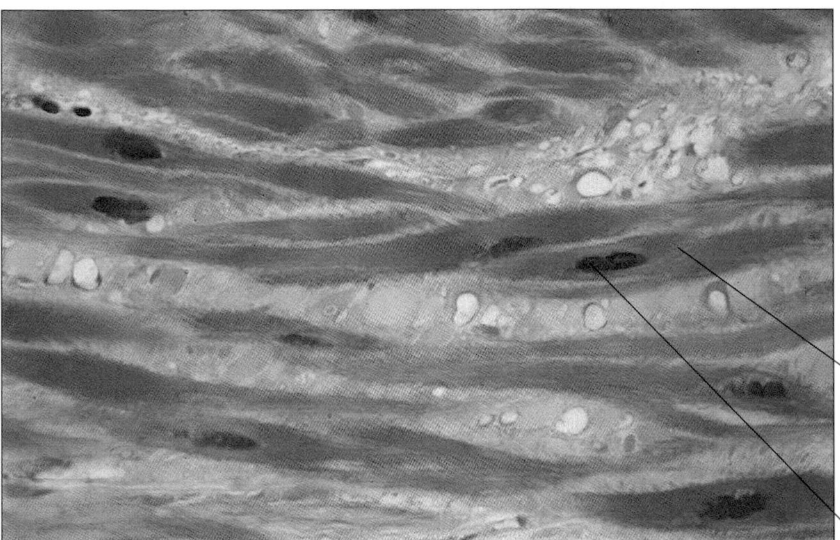

Figure 1-42
Smooth Muscle (Longitudinal Section)
Canoe- or spindle-shaped muscle cells lack striations, and each has a single, elongated nucleus. (×250)

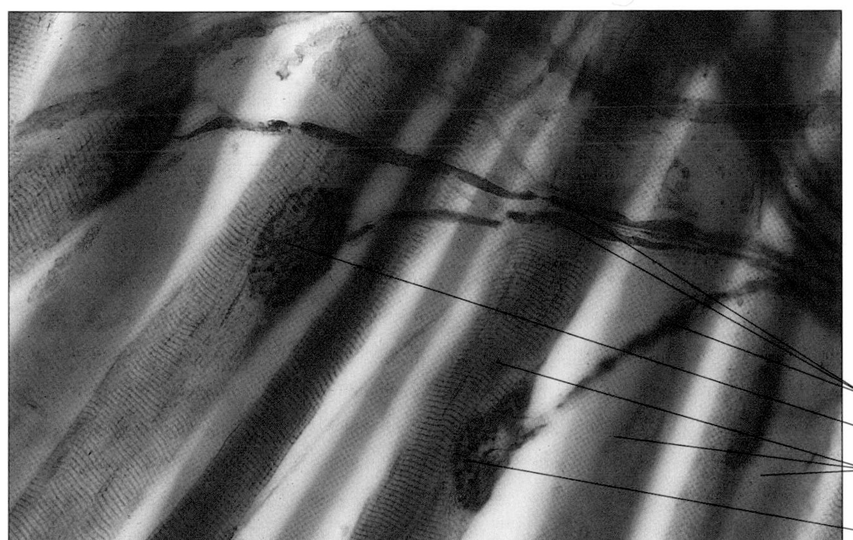

Figure 1-43
Innervation of Skeletal Muscle: Motor Endplate Branching nerve bundle terminates to form the **myoneural junctions.** Nerve terminals release small quantities of chemical neurotransmitter to stimulate muscle contraction.

Figure 1-44
Astrocytes (Neuroglia) Star-shaped supporting cells of central nervous system modulate ionic environment. Cytoplasmic extensions make contact with blood vessel. Cat. (Silver stain; ×280)

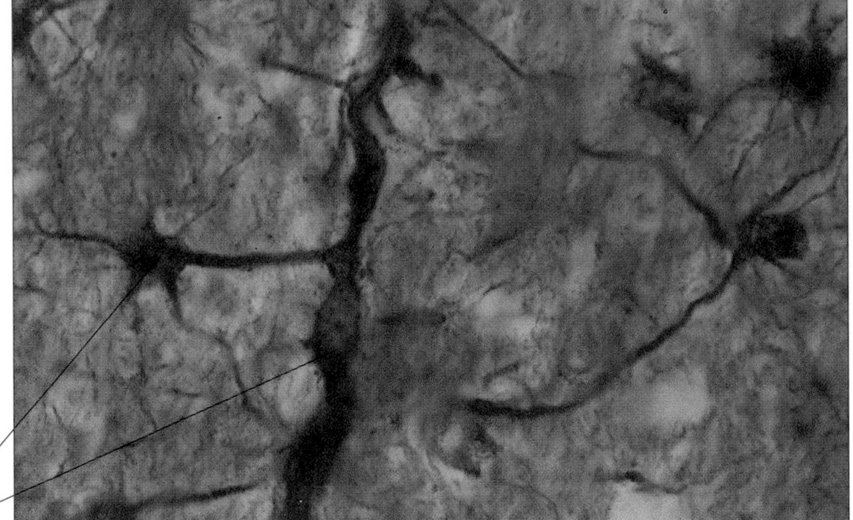

Figure 1-45
Purkinje Cells (Neurons) Numerous branched processes (dendrites) receive information for processing. Single process (axon) sends information to other neurons. Human cerebellum. (×100)

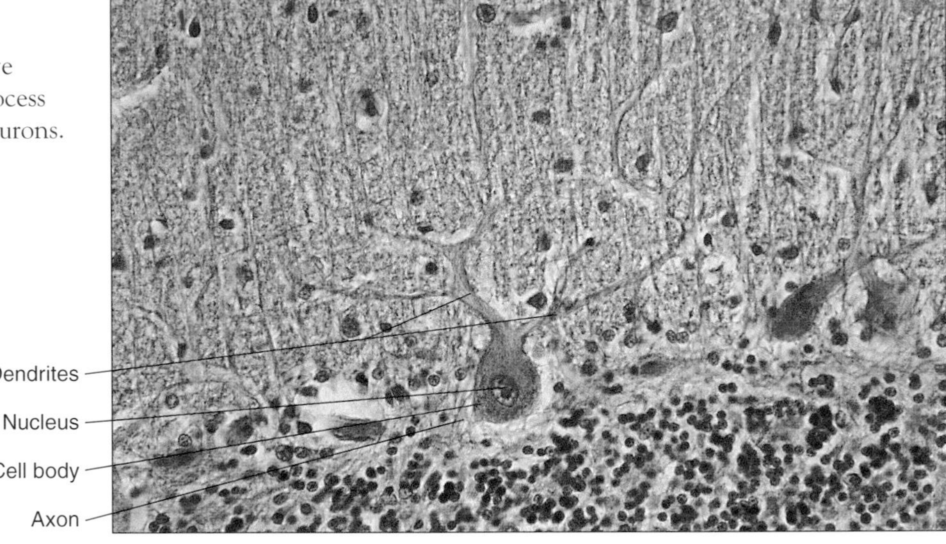

Figure 1-46
Pyramidal Cells Neurons from human cerebral cortex directly receive information from hundreds of other cells; send information on to hundreds of others. (Fox-Golgi stain; ×100)

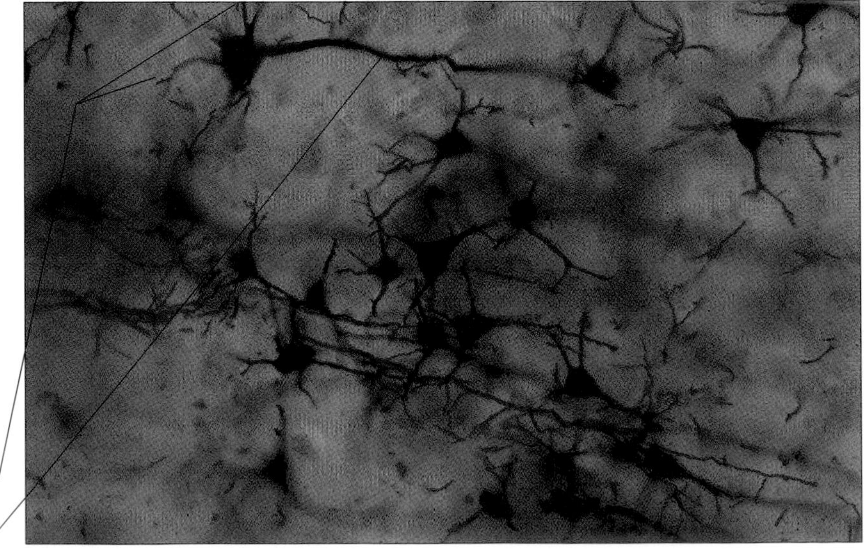

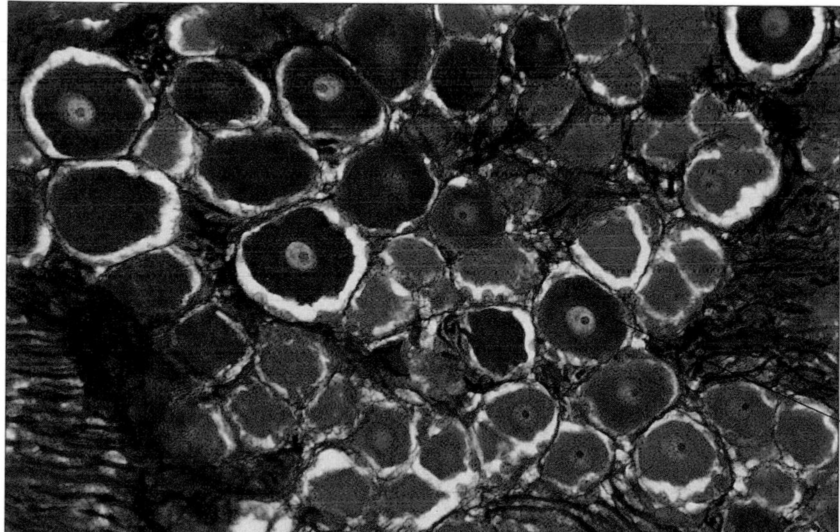

Figure 1-47
Dorsal Root Ganglion Sensory signals representing pain, temperature, pressure, muscle tension, joint position, and others depend on these cells. Their dendrites collect sensory information throughout the body and axons route it into the spinal cord. (×100)

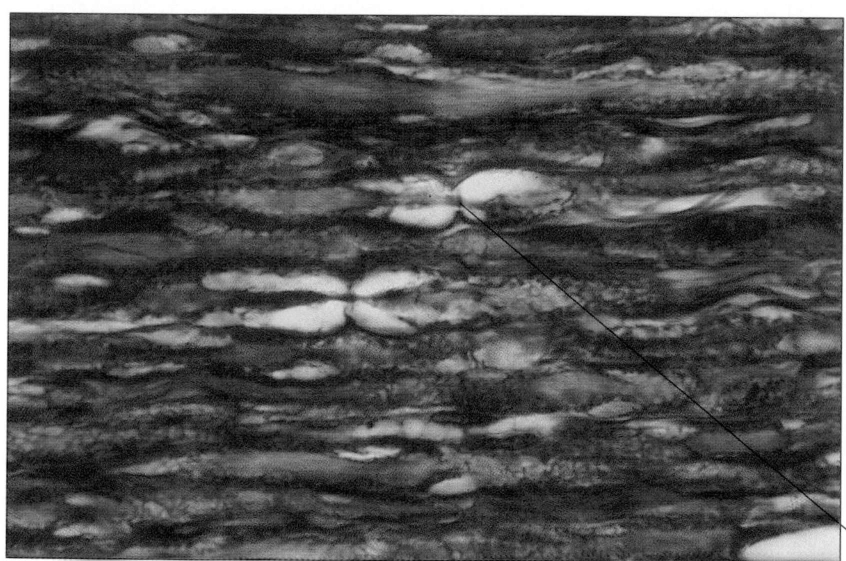

Figure 1-48
Nerve Fibers (Longitudinal Section) Clear areas show dimpling characteristic of nodes of Ranvier. (×250)

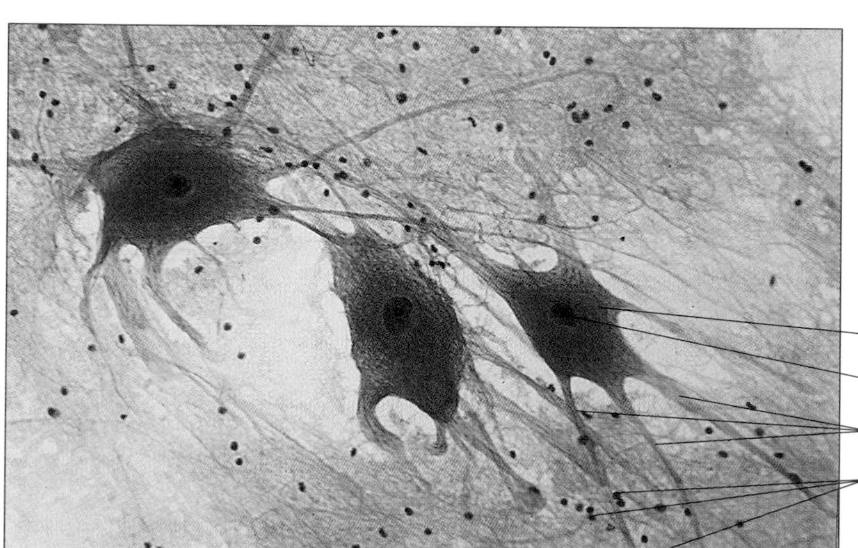

Figure 1-49
Motor Neurons of the Spinal Cord
Integrated command information from the brain and sensory signals enter these cells, whose efferent activity controls muscular contraction. Numerous synapses occur on dendrites and cell body (soma). (×50)

Figure 1-50
Myelinated Nerve Fibers (Cross Section)
Central core stains dark; insulating myelin appears white. (×250)

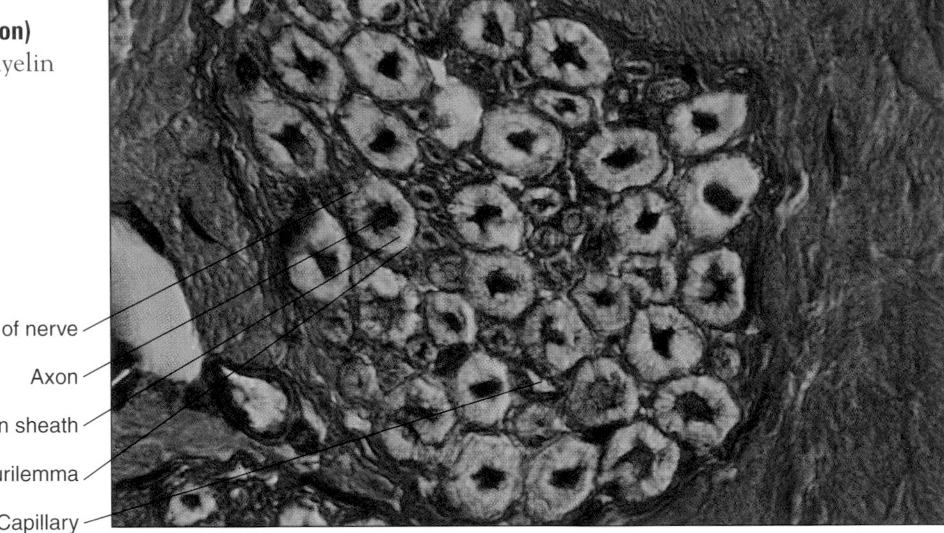

Figure 1-51
Spinal Cord, Lumbar Region (Cross Section)
Top is dorsal, bottom is ventral. Light central dot is central canal. Darkly staining H-shaped region is gray matter of cell bodies; surrounding lighter material is composed of myelinated axons. Ventral horns of gray matter contain motor neurons; dorsal horns contain cell bodies of sensory pathways. (×4)

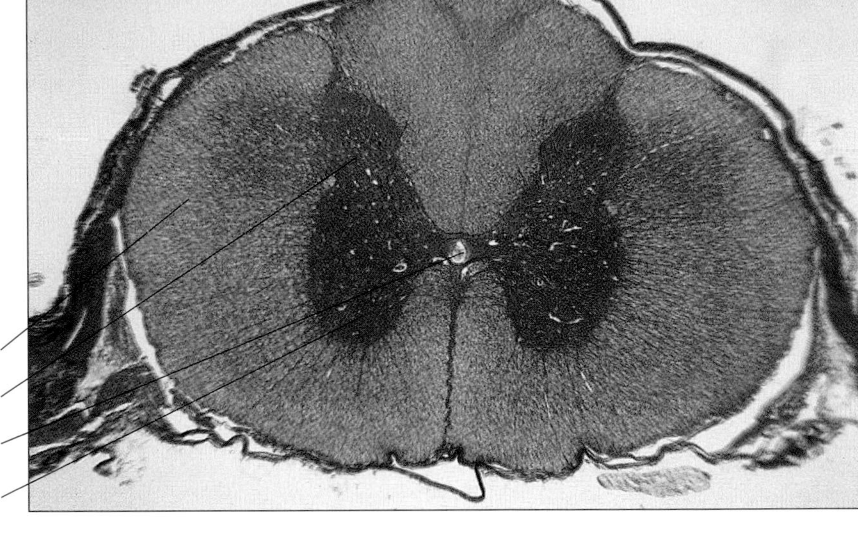

Figure 1-52
Retina Layered structure evident. Dark line of cells near top is pigment epithelium. Broad striped region represents photoreceptors (rods and cones), whose nuclei stain heavily immediately beneath. Below receptor nuclei lie synaptic region and a layer of nuclei belonging to bipolar cells. Bipolar cell output synapses onto ganglion cells, only a few of which appear near bottom. Axons of ganglion cells form optic nerve. (×100)

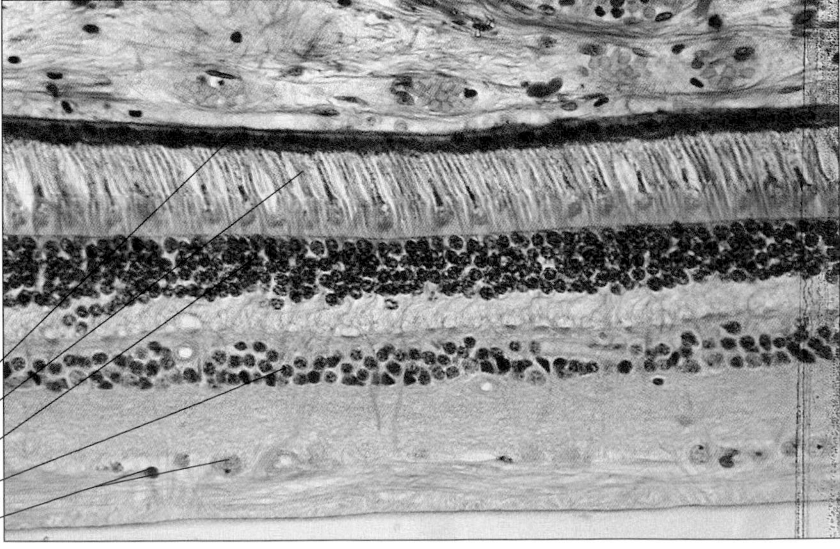

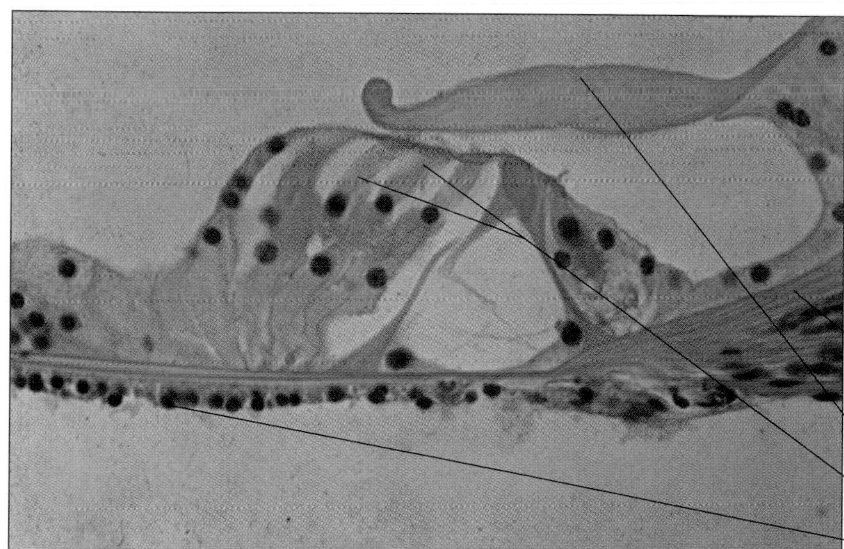

Figure 1-53
The Organ of Corti Thick finger of tectonial membrane extends from right to stimulate complex of four hair cells (three on left, one on right) of central structure that rests on basilar membrane. Nerve fibers from hair cells exit right to spiral ganglion for processing and transmission of messages to brain. (×500)

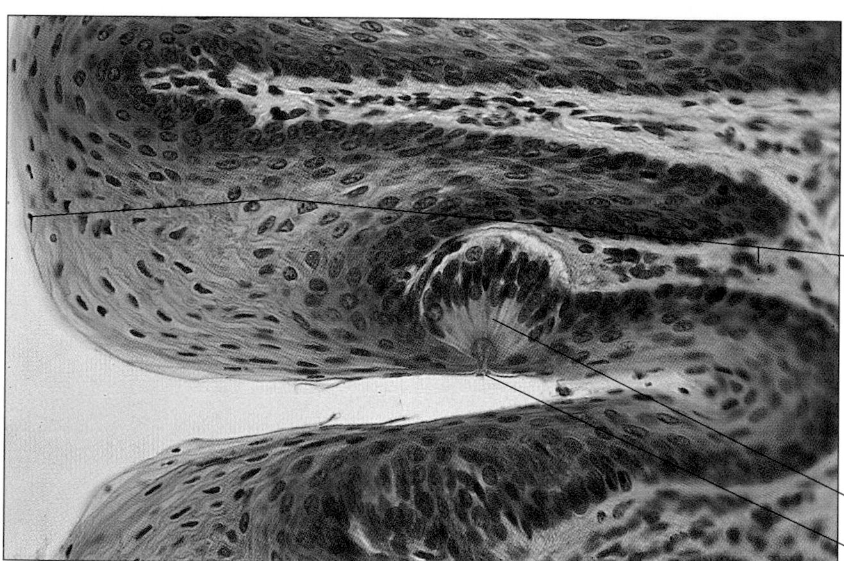

Figure 1-54
Taste Bud Dissolved chemicals enter fungiform papilla through small pore to directly stimulate sensory cells and initiate taste perception. (×100)

Figure 1-55
Thyroid Gland Follicles Cuboidal epithelium surrounds endocrine follicles of the thyroid gland, the only gland that stores substantial amounts of its own hormone. (×100)

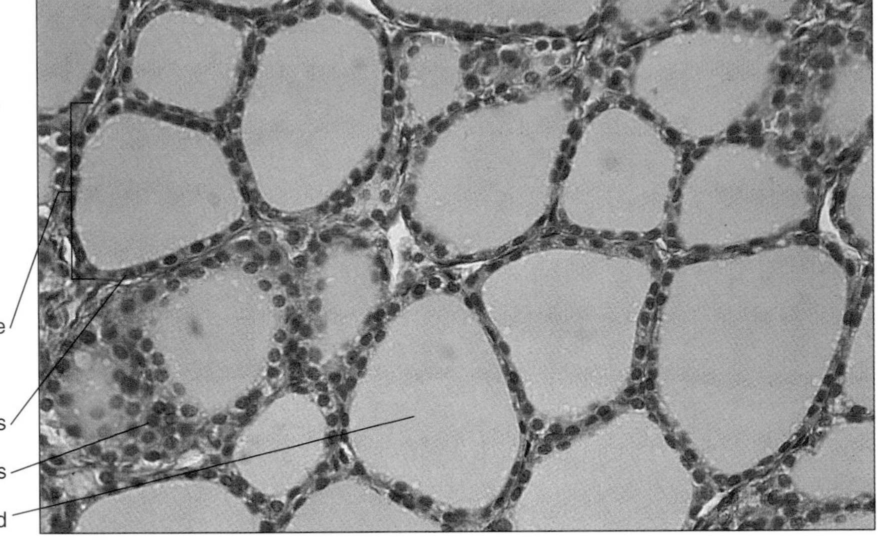

Figure 1-56
Parathyroid Gland Section through the parathyroid gland. (×40)

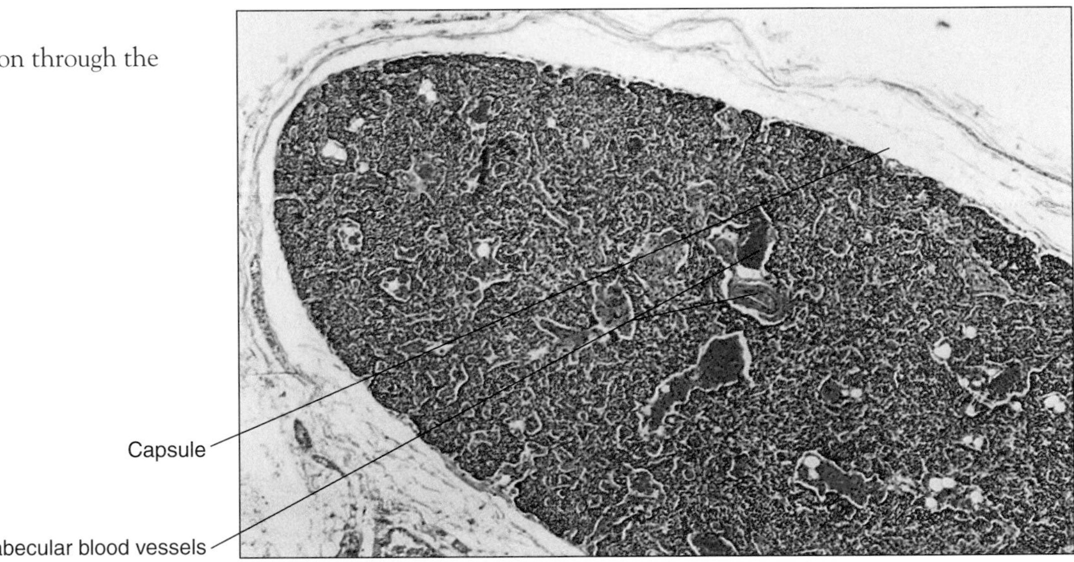

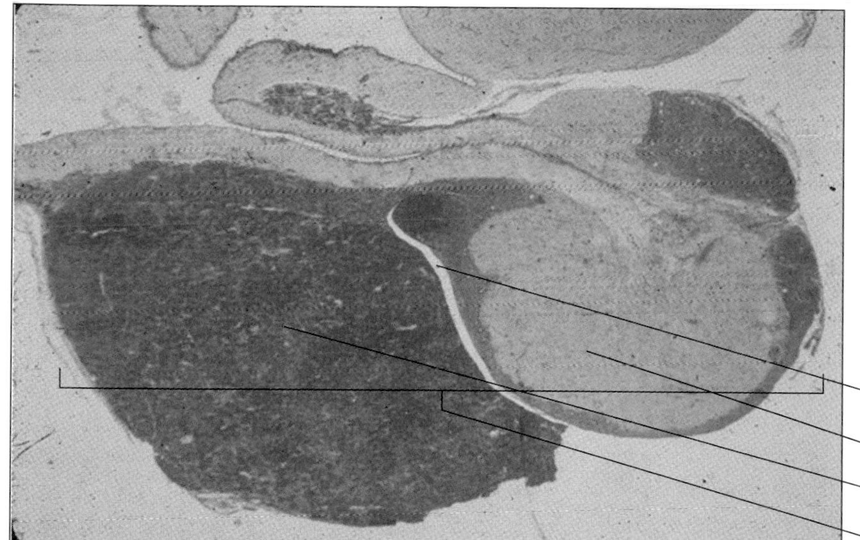

Figure 1-57a
Pituitary Gland The pituitary gland consists of two components: the posterior component, or neurohypophysis (light stain), consists of mainly nervous tissue, whereas the anterior component, or adenohypophysis (dark stain) consists of a glandular epithelium. (×10)

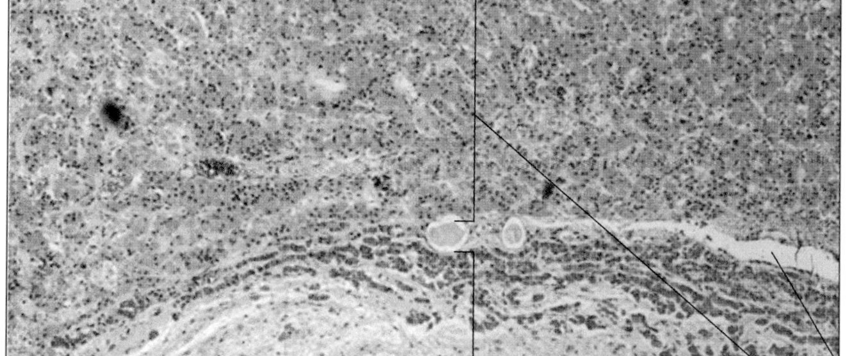

Figure 1-57b
Pituitary Gland The cleft between the neurohypophysis and adenohypophysis is visible in this view of the pituitary gland. (×100)

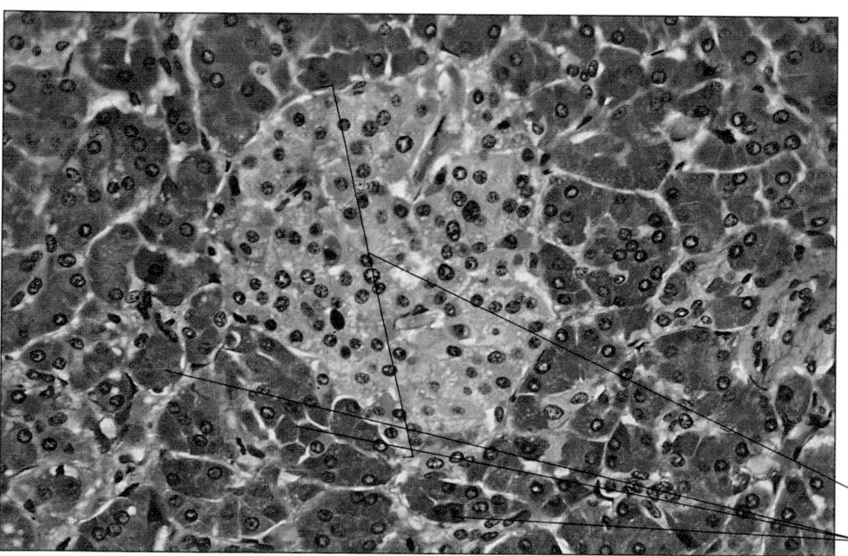

Figure 1-58
Pancreas The pancreatic islet of Langerhans cells form the endocrine portion of the pancreas. Alpha cells secrete glucagon, beta cells secrete insulin, and delta cells secrete somatostatin. The exocrine portion of the pancreas secretes digestive enzymes through a series of ducts.

Figure 1-59
Adrenal Cortex Outer zone of rounded groups of cells (zona glomerulosa) secretes mineralcorticosteroids (aldosterone). Middle zone of cells appearing in rows (zona fasciculata) secretes glucocorticosteroids. Innermost zone of cells arranged in a meshwork (zona reticularis) secretes mainly androgens. (×50)

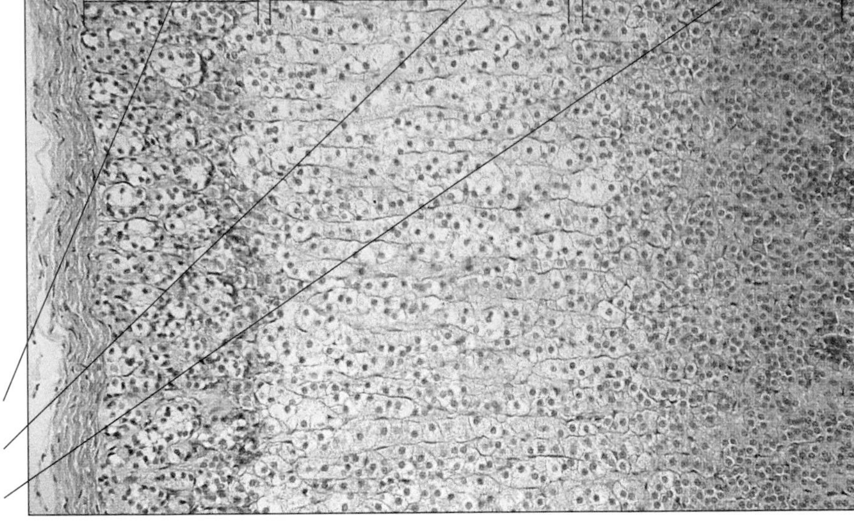

Figure 1-60
Neutrophil Most numerous (65%) of the leukocytes, it is characterized by a multilobed nucleus and granular cytoplasm. Engages in phagocytosis. (Neutral dyes stain; ×640)

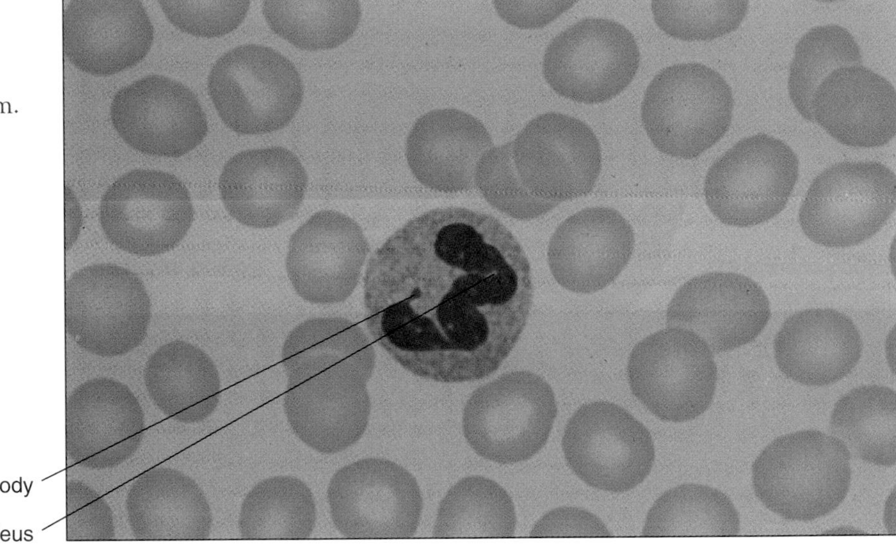

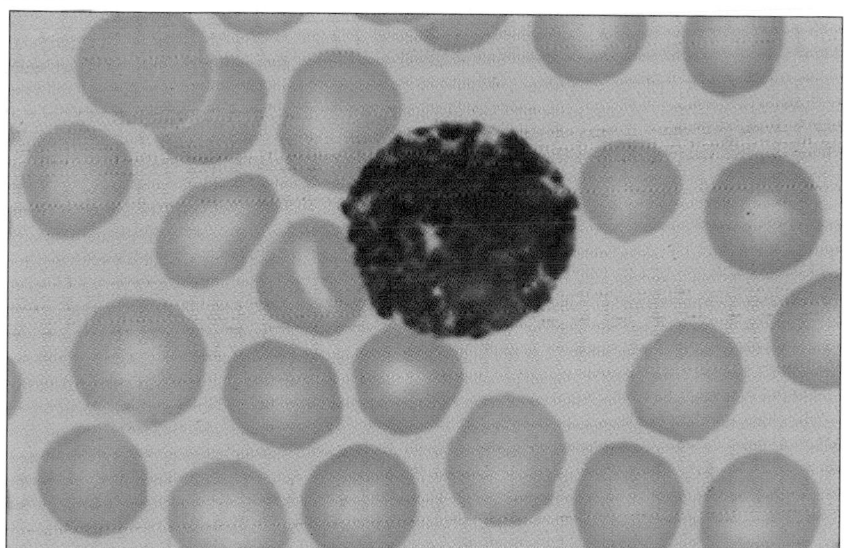

Figure 1-61
Basophil Normally the rarest (1%) of the leukocytes, its kidney-shaped nucleus may be almost obscured by cytoplasmic granules. These cells contain numerous chemicals involved in inflammation. (Basic dyes stain; ×640)

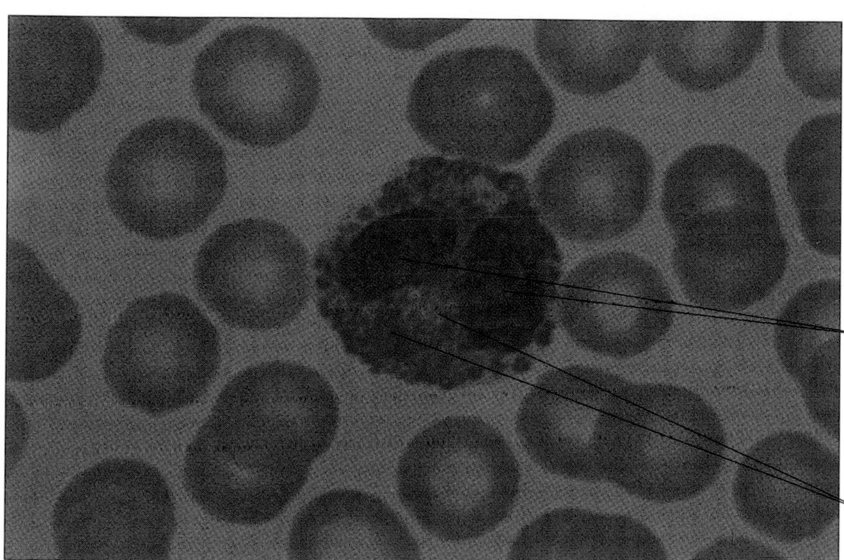

Figure 1-62
Eosinophil Relatively rare (6%) leukocyte. Usually identifiable because of red-to-orange-staining cytoplasmic granules. Function not definitely known but elevated especially in allergies. (Selective eosin stain; ×640)

Figure 1-63
Lymphocyte Common (25%). Characterized by single-lobed, "dented" nucleus surrounded by clear cytoplasm. May be large or small. Heavily involved in the immune response including synthesis of antibodies. (×640)

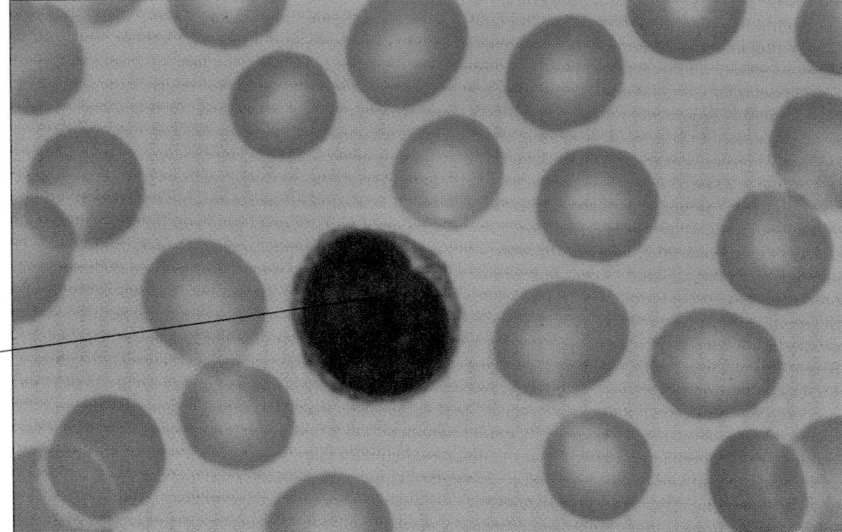

Figure 1-64
Monocyte Relatively rare (3%). Lobed, often kidney-shaped nucleus is surrounded by clear cytoplasm. Largest of the leukocytes, this cell is a precursor to a macrophage, which engages in phagocytosis. (×640)

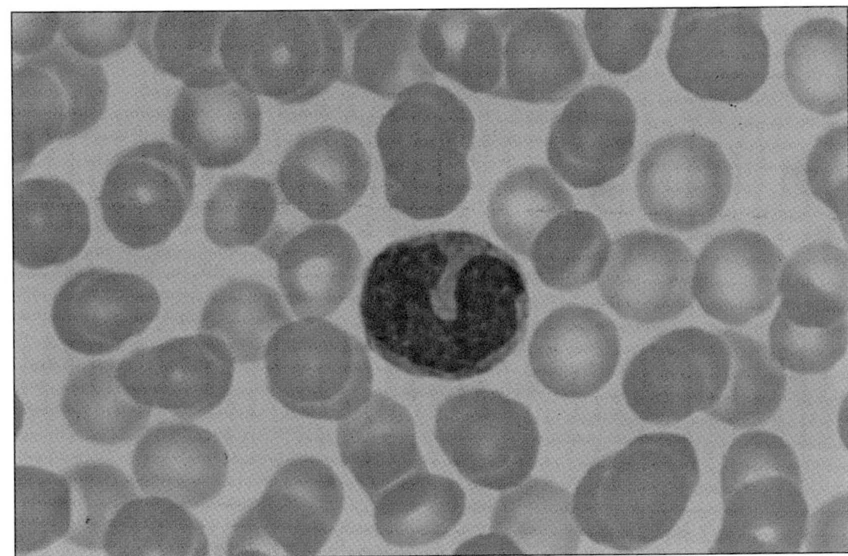

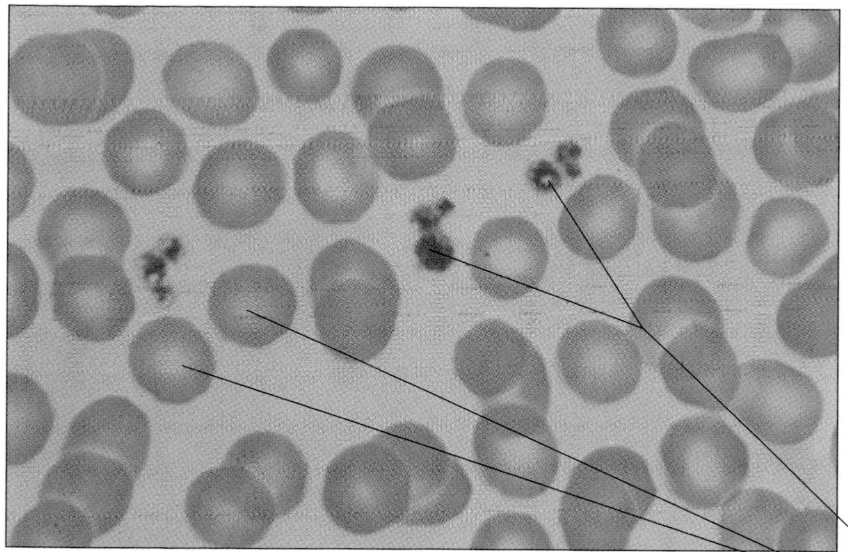

Figure 1-65
Erythrocytes (Red Blood Cells) and Platelets Circulating erythrocytes are far more common than any of the leukocytes. Normally they have no nucleus but contain the red pigment hemoglobin, which permits them to transport oxygen and carbon dioxide throughout the body. Typically they assume the shape of a biconcave disk. Their diameter of about 7 microns is useful for comparing sizes of other histological structures. Platelets are cellular remnants of a much larger precursor. These remnants contain numerous chemicals, including those important for clotting and inflammation. Platelets initiate blood clotting by forming a plug at wound sites. (×500)

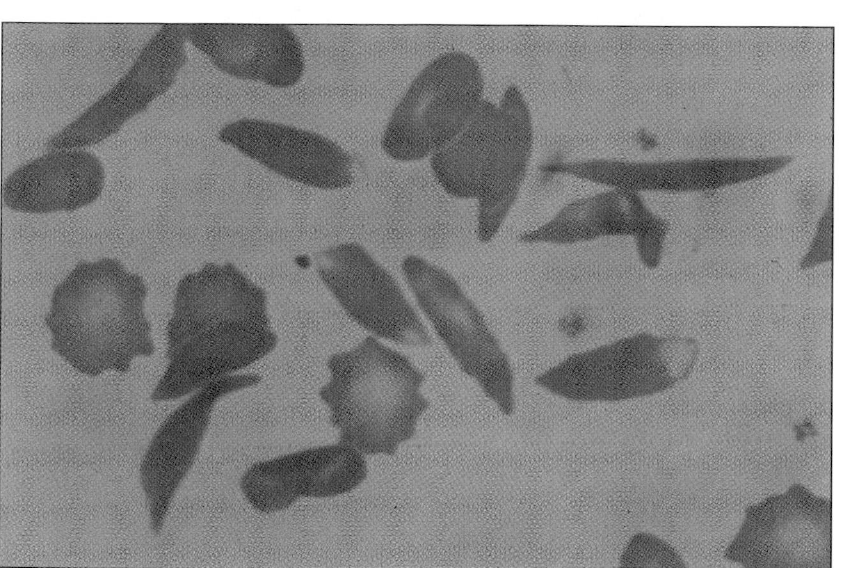

Figure 1-66
Sickle Cell Anemia Genetic alteration of hemoglobin results in altered membrane structure and abnormal wavy or elongated, curved shape that often resembles a sickle *(upper left)*. Oxygen-carrying capacity is much reduced. (×500)

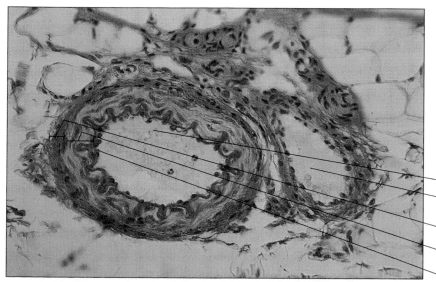

A
V
Tunica adventitia
Tunica media
Tunica intima

Figure 1-67
Artery (A) and Vein (V) Blood vessels possess a **tunica intima** that lines the lumen, outside of which is a muscular **tunica media,** and a connective tissue covering, the **tunica adventitia.** The tunica media of arteries is typically much thicker than that of veins. (×100)

Figure 1-68a
Arterial Cross Section Single layer of darkly stained cells, the tunica intima lines the lumen. Thick tunica media is composed of canoe-shaped smooth muscle cells. Outer adventitial layer of connective tissue provides elastic support and strength. (×50)

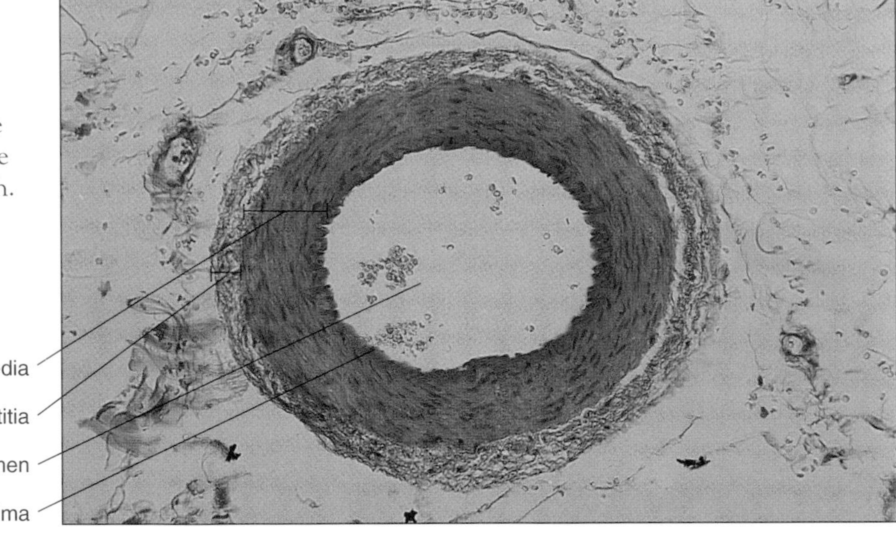

Figure 1-68b
Atherosclerosis Cross section of an artery with advanced atherosclerosis. (×40)

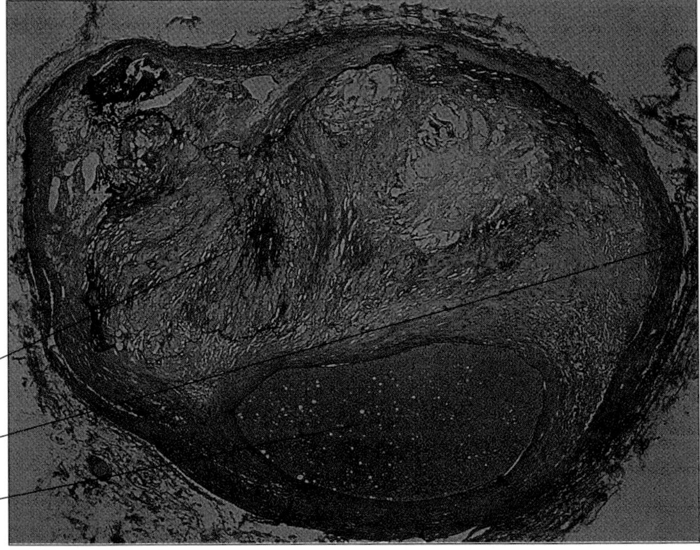

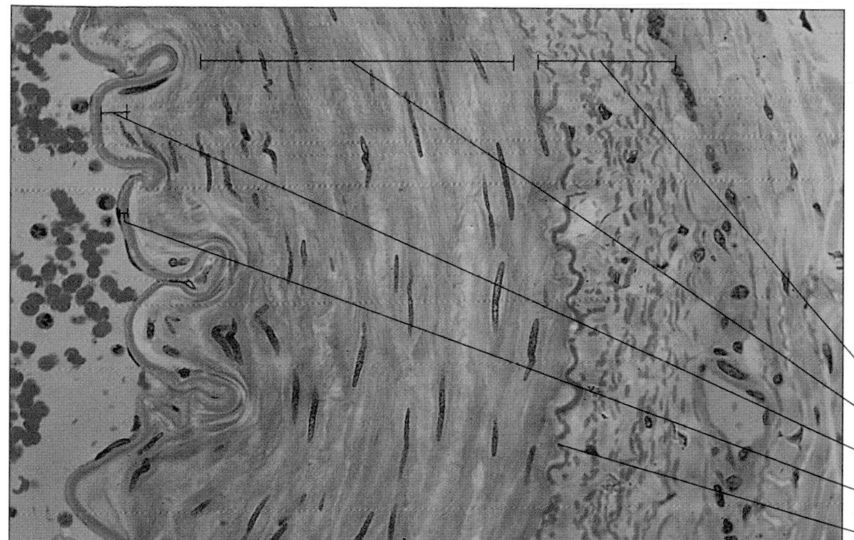

Figure 1-69
Detail of Arterial Wall Inner endothelial cells of tunica intima *(left)* lie on a basement membrane. A thin layer of smooth muscle cells and elastic tissue (lamina propria) throws this tunic into folds. The tunica media contains multiple layers of smooth muscle cells regularly arranged. A wavy external elastic membrane separates the tunica media from the adventitia.

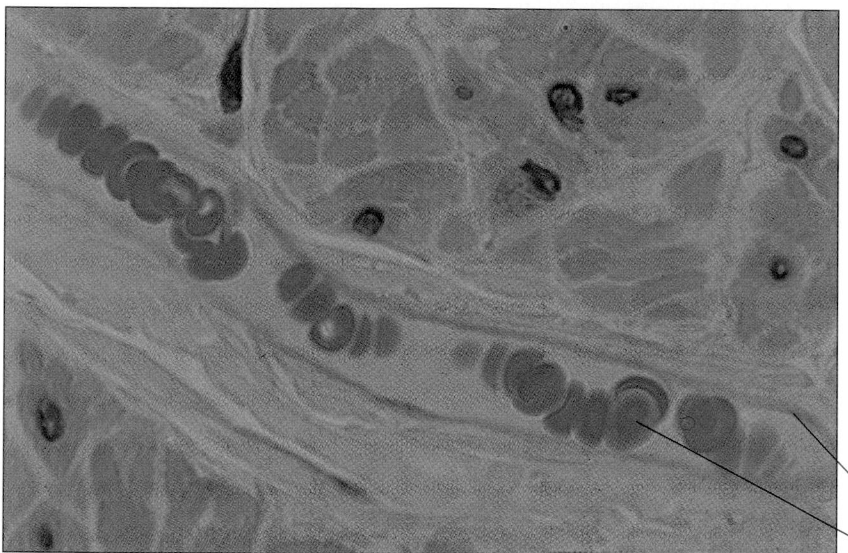

Figure 1-70
Capillary with Red Blood Cells in Single File Capillary wall is made of flattened endothelial cells without complex tunics, a simple structure that facilitates the exchange of gases, nutrients, wastes, and hormones. (×400)

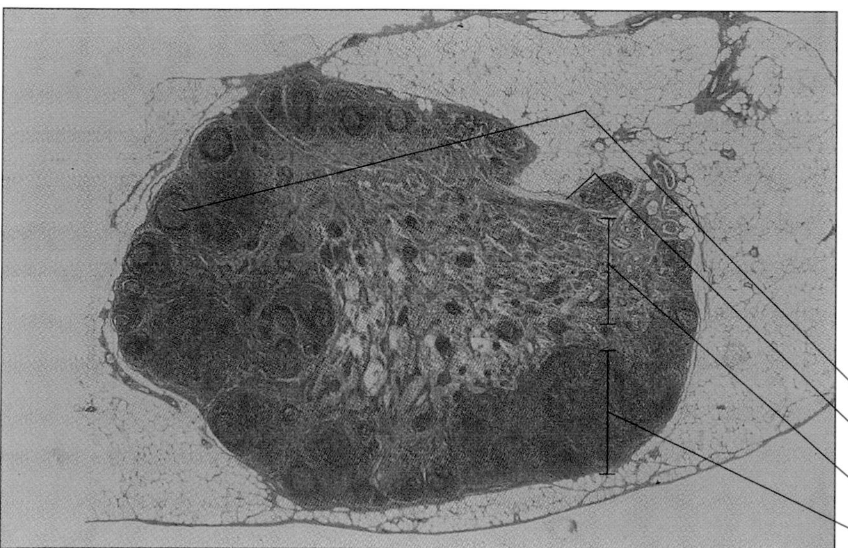

Figure 1-71
Lymph Node Outer cortex containing several follicles surrounds medulla, with its narrow, dark medullary cords. Notch is the hilum, through which blood and lymphatic vessels pass. (×5)

Figure 1-72
Valve of Lymphatic Vessel One-way flow of lymph, from left to right in this figure, is ensured by valve action in lymph vessel. Vessels themselves are thin walled and lack musculature; pumping action occurs through compression by neighboring muscles. (×25)

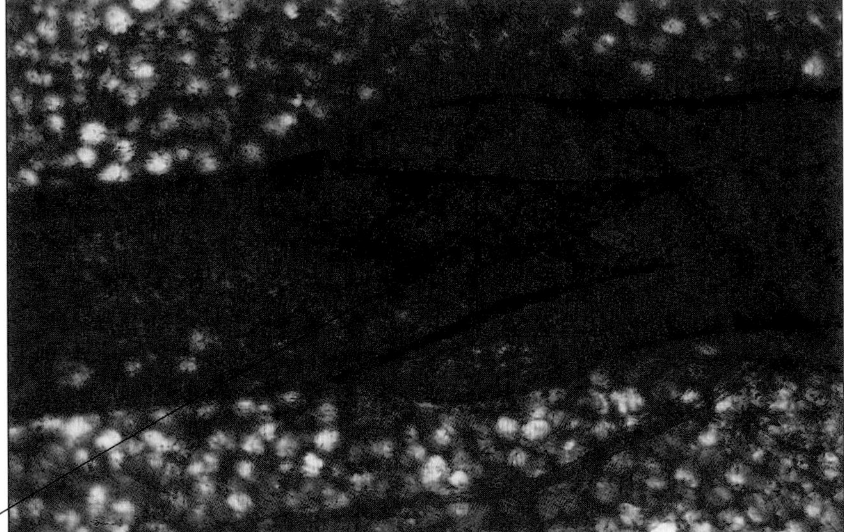

Figure 1-73a
Thymus Various lobules contain thick, darkly staining cortex surrounding a smaller, lighter-staining medulla. Small, round cellular patches in medulla are Hassall's corpuscles. In adults, much of thymus degenerates and is replaced by adipose tissue. (×10)

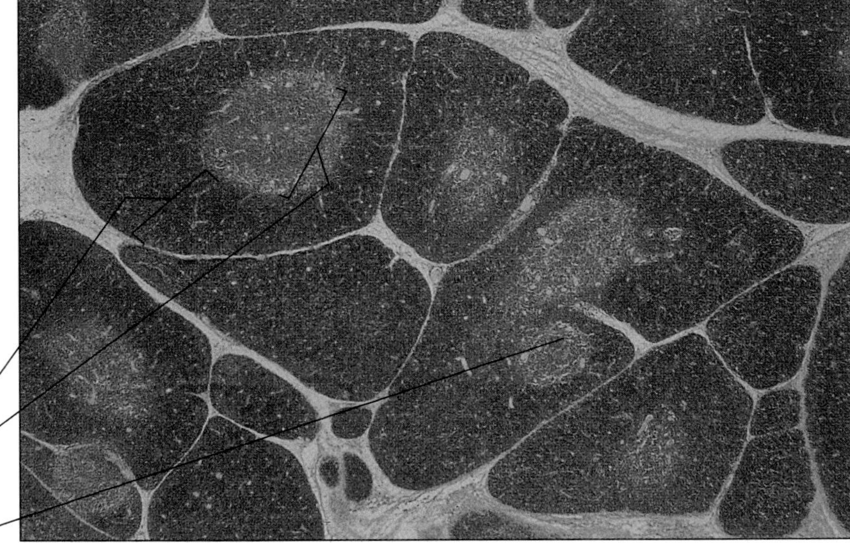

Figure 1-73b
Thymus Under higher magnification, the appearance of Hassall's corpuscles distinguish the thymus from other organs. Surrounding the corpuscles are reticulate epithelial cells. (×400)

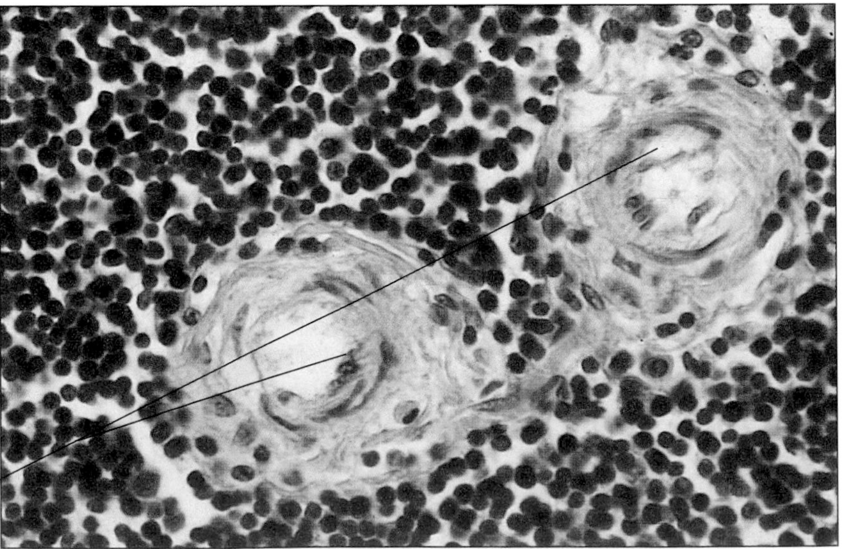

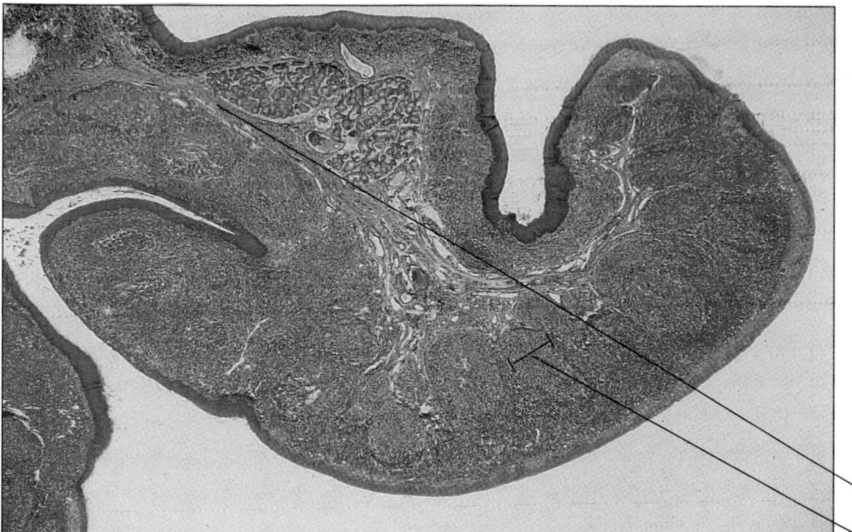

Figure 1-74
Palatine Tonsil Outer capsule surrounds subcapsular sinus, under which are several large, rounded germinal centers surrounding trabecular arteries and veins. Efferent lymph vessel leads out to upper left. (×5)

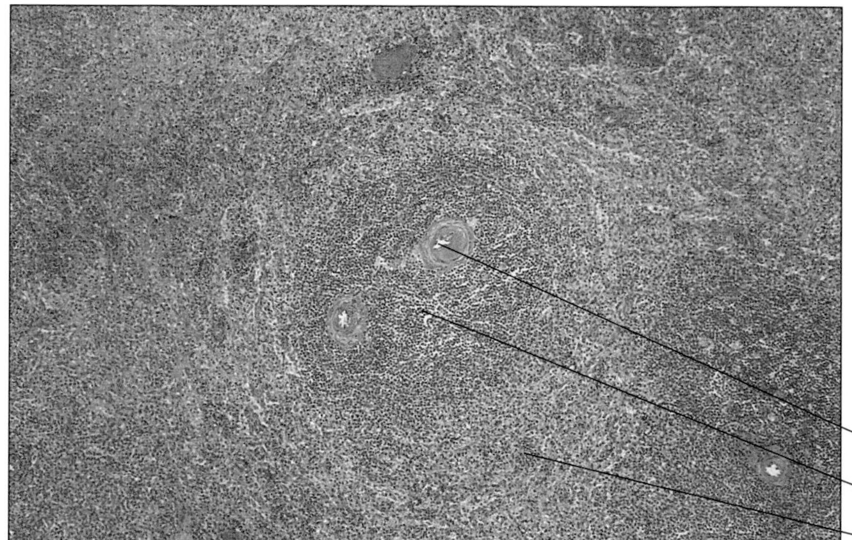

Figure 1-75
Spleen Central blood vessels are surrounded by area of densely staining white pulp composed of lymphoid cells. Less densely staining red pulp, with fewer cell nuclei, surrounds white pulp. (×25)

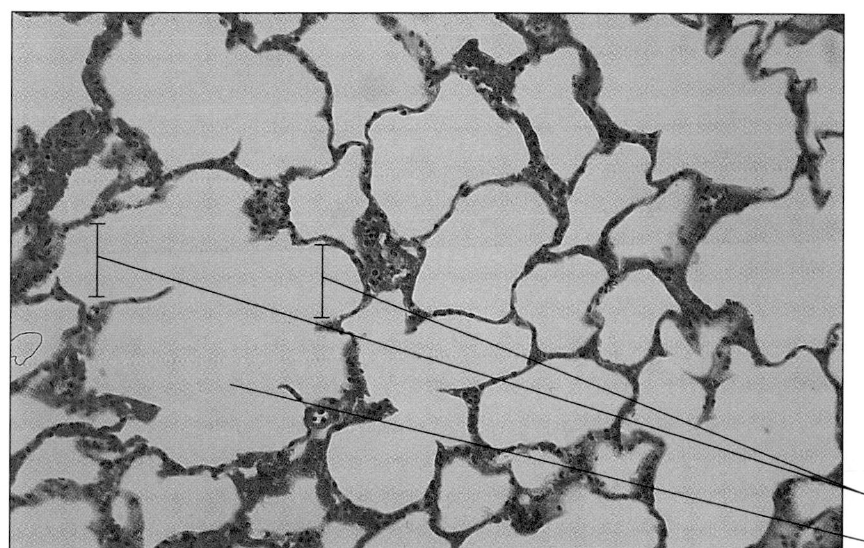

Figure 1-76
Alveoli Thin-walled respiratory exchange surfaces aid in rapid diffusion of gases. Bronchiole terminates at atrium, which acts as entryway into several individual alveoli, greatly multiplying surface area. (×50)

Figure 1-77
Details of Alveolus Squamous cells compose alveolar wall, which is bordered by thin-walled blood vessels *(upper left)* containing erythrocytes. (×100)

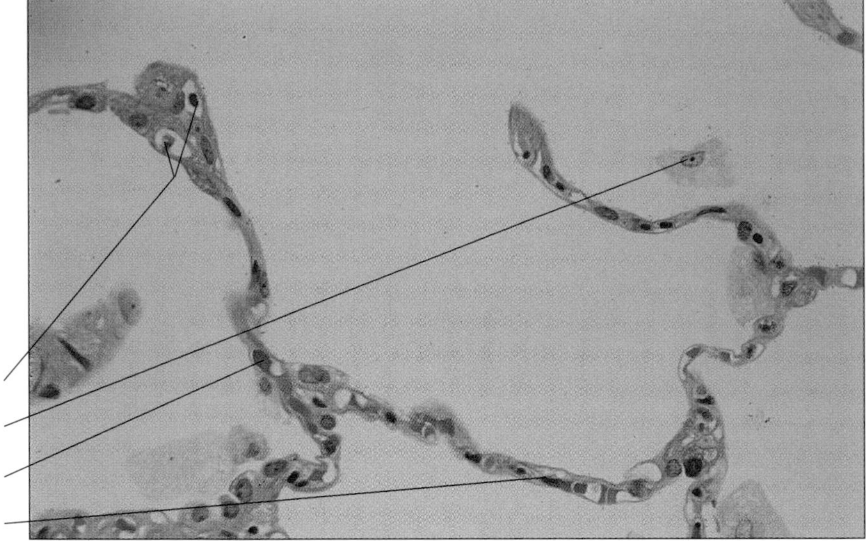

Figure 1-78
Bronchiole Epithelial layer that lines the lumen is surrounded by layer of smooth muscle, which regulates bronchiolar diameter. Round structures outside of smooth muscle layer are blood vessels. (×100)

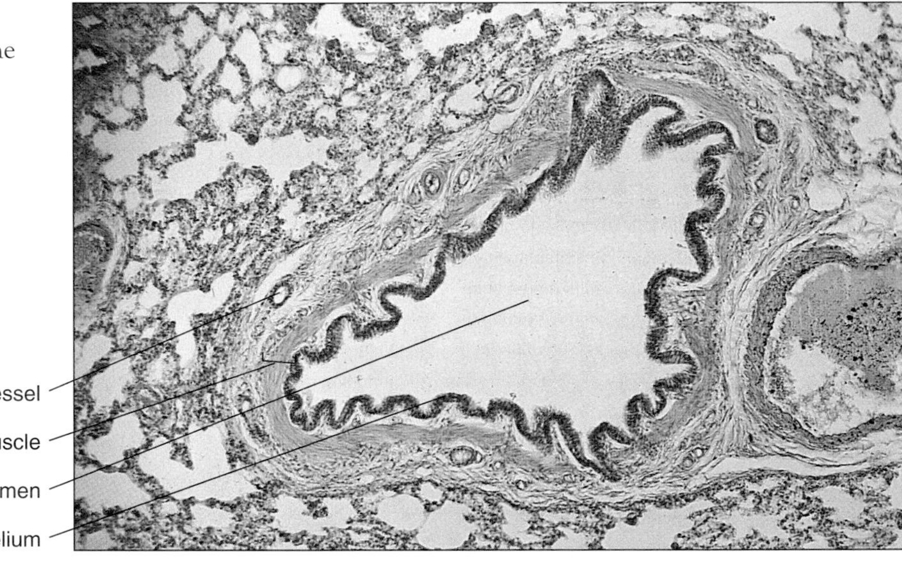

Figure 1-79
Esophagus Surrounding the lumen, esophageal structure contains, in order, the four basic layers of the alimentary canal: **mucosa** (composed of epithelium, the thick lamina propria, and dark muscularis), **submucosa** (light with spaces, blood vessels, and lymph channels), two thick layers of the **muscularis** (circular and longitudinal), and the thin, connective **adventitia** on the surface. Cross section, human. (×3)

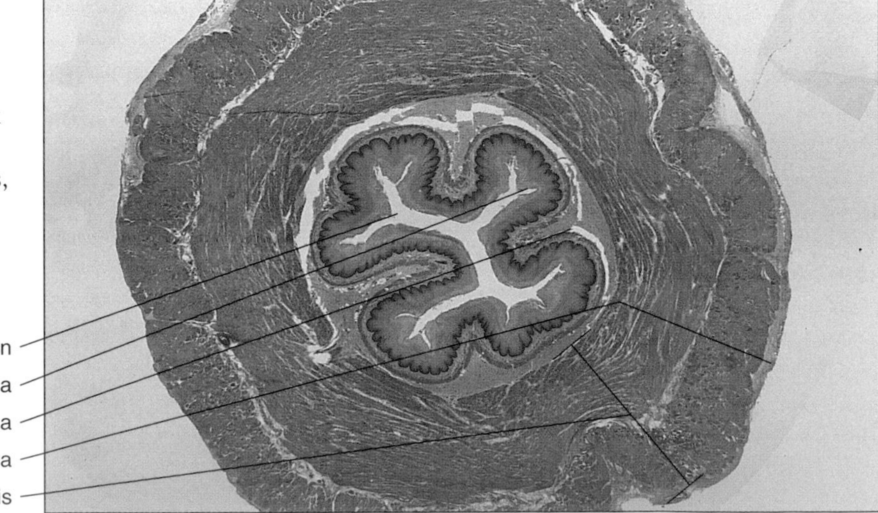

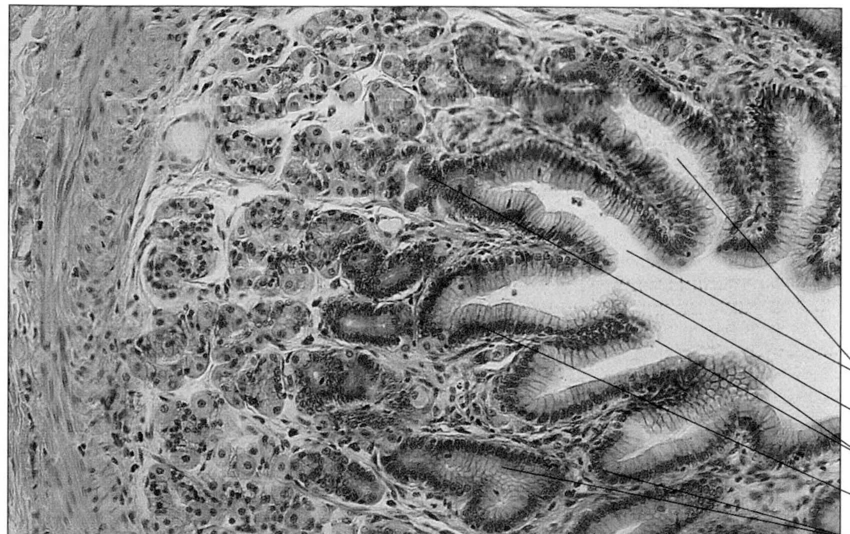

Figure 1-80a
Stomach Mucosa Visible at entrances to gastric pits are mucus-secreting goblet cells of columnar epithelium. Deeper in pits are acid-secreting parietal cells and enzyme-secreting chief cells. Endocrine-secreting cells near tip of pits are noncolumnar and smaller, with dark, round nuclei. Gastric pits penetrate deep into submucosal layer. Edge of muscularis layer is visible. (×50)

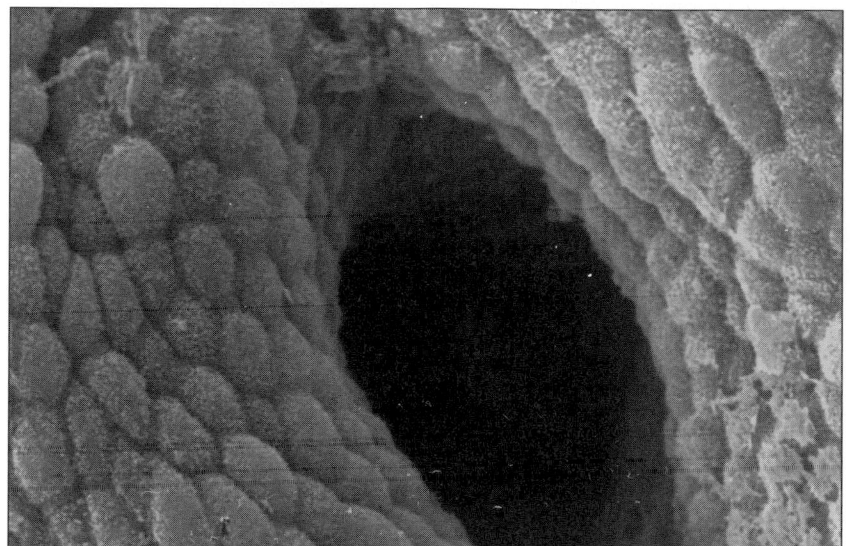

Figure 1-80b
Gastric Pit The opening of a gastric pit into the stomach, surrounded by the rounded apical surfaces of the columnar epithelial cells of the mucosa.

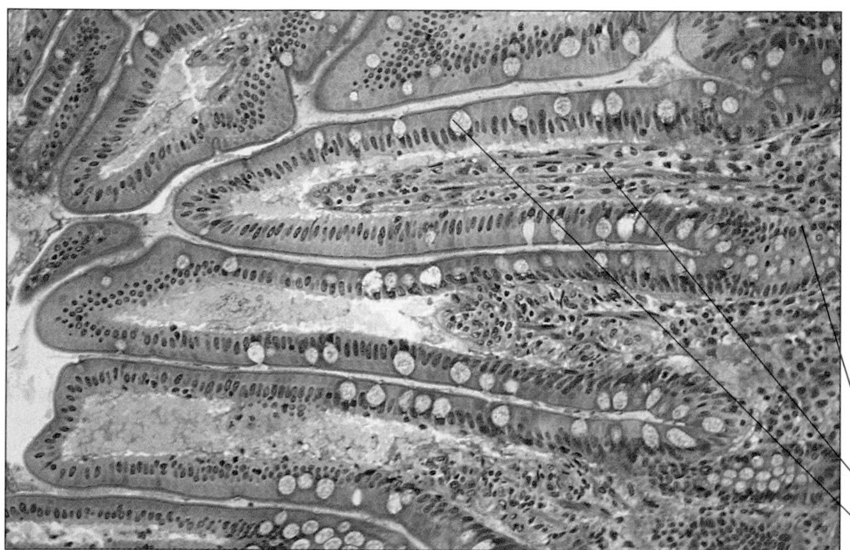

Figure 1-81
Small Intestine, Villi of Ileum (Longitudinal Section) Numerous pale goblet cells punctuate columnar epithelium that covers each villus. Core of villus contains small blood vessels and blind lymph channel (lacteal). Deep in crypts are endocrine cells, identifiable as dark, round nuclei in a noncolumnar cytoplasm. Human. (×50)

Figure 1-82
Small Intestine, Villi of Ileum (Cross Section)
Goblet cells emptying contents through brush border surface are evident. Core of villus contains blood vessels, lymph channels, and lymphocytes. Human. (×100)

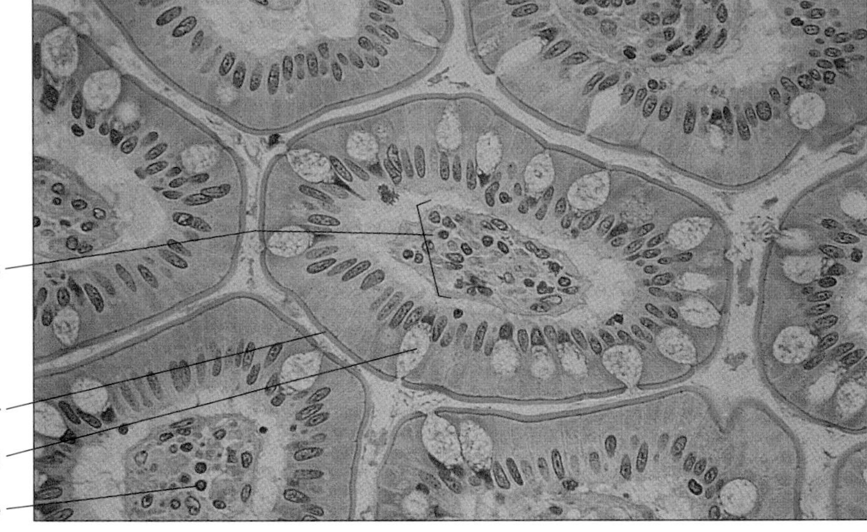

Figure 1-83
Large Intestine (Colon) (Cross Section)
Surface is thrown into folds but devoid of villi. Thick submucosa contains blood vessels and lymph channels. (×10)

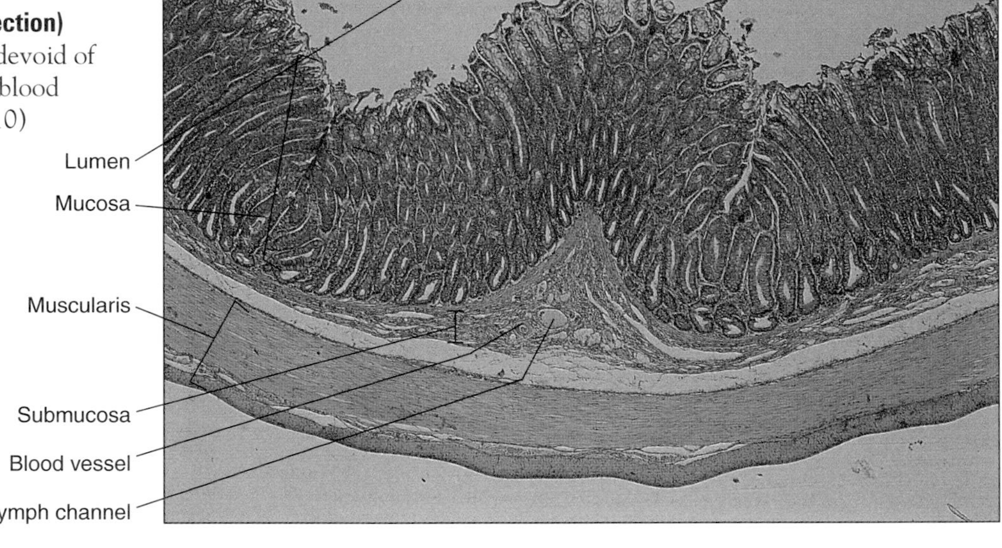

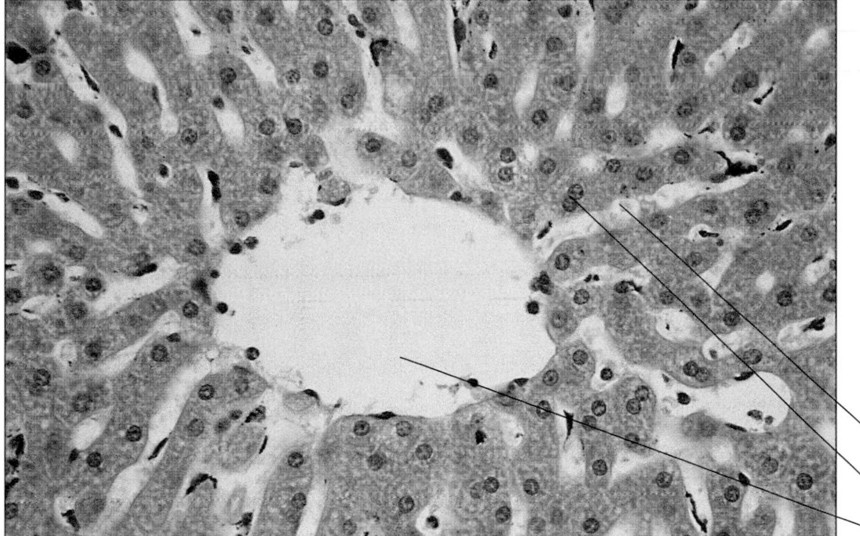

Figure 1-84a
Liver with Central Vein and Sinusoids Parenchymal hepatocytes lie in radial arrangement around central vein that is lined with single endothelial layer. Cords of hepatocytes are separated by spaces (sinusoids). Sinusoidal surface is covered by microvilli. (×100)

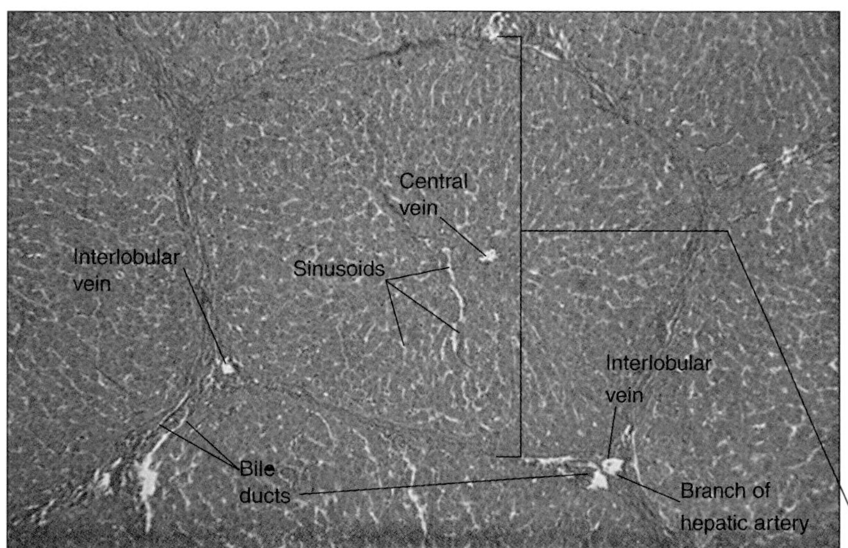

Figure 1-84b
Liver The liver consists of numerous lobules. A single lobule is in the center of view. At the junction of three adjacent lobes is a bile duct, a branch of the hepatic artery, and a branch of the hepatic portal vein. These three tubes are called a triad. (×40)

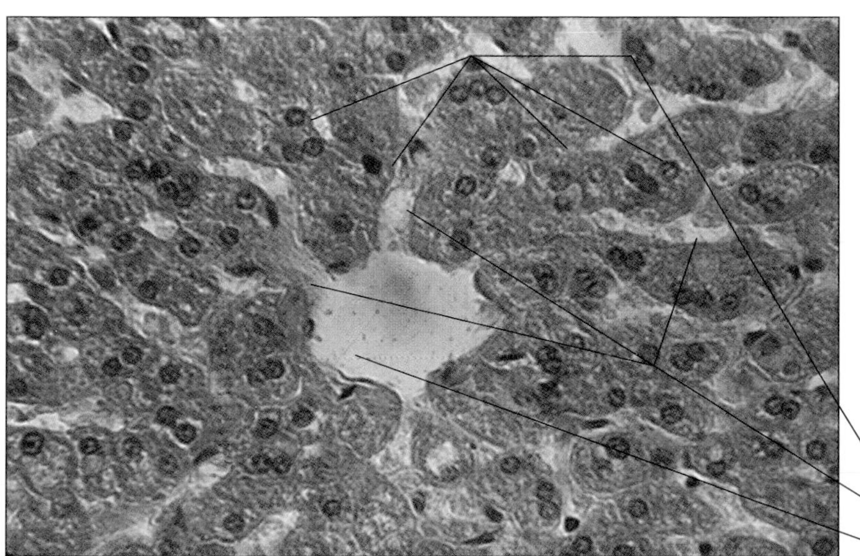

Figure 1-84c
Liver A single liver lobule consists of a central vein (shown in the center), which collects blood as it flows through narrow endothelial-lined channels, or sinusoids. The cells bordering the sinusoids are called hepatocytes. (×400)

Figure 1-85
Gallbladder Mucosal folds are covered by epithelium with well-developed microvilli. Lamina propria contains blood vessels. (×25)

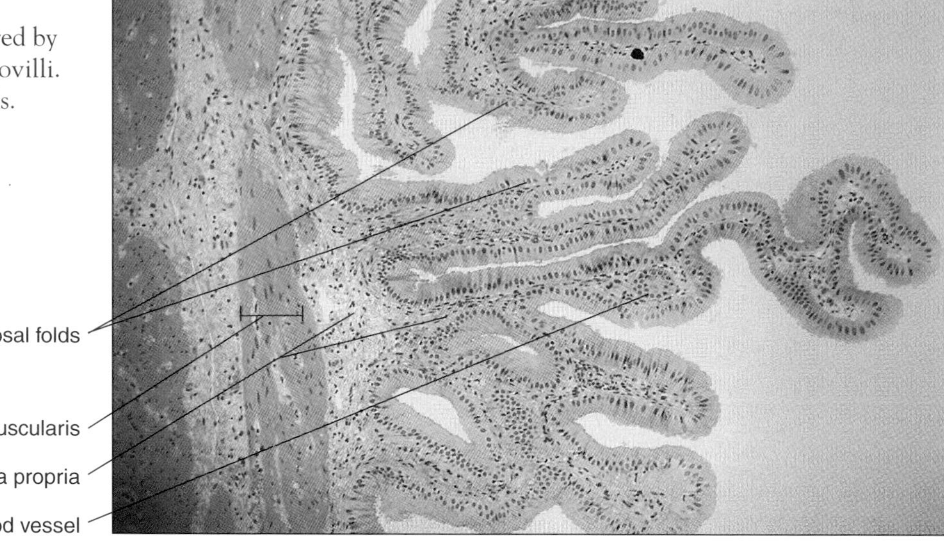

Figure 1-86
Vermiform Appendix (Cross Section)
Overall structure resembles that of colon. Large, darkly staining structures are lymphoid follicles, the size and number of which decrease with age. Human. (×3)

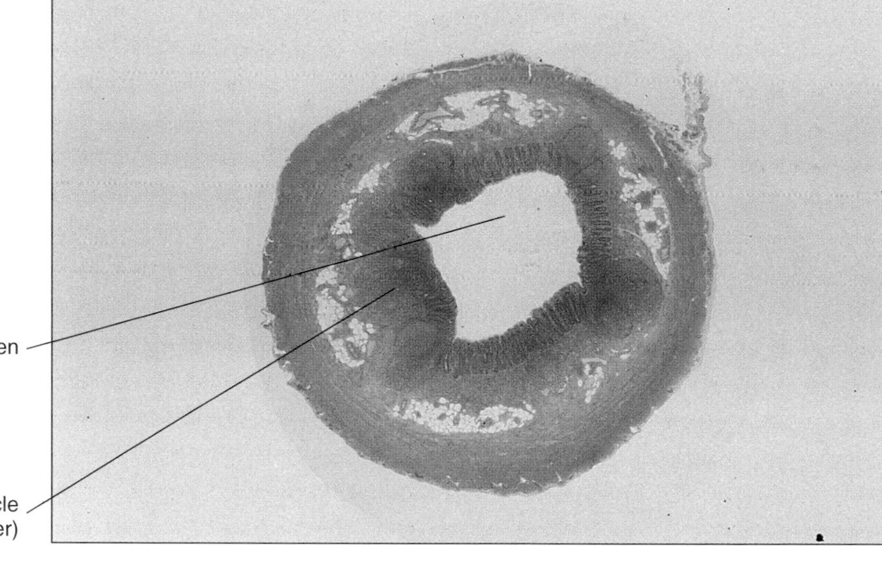

Figure 1-87
Sublingual Salivary Gland Large, pale, mucus-secreting cells, some with caps of serous demilunes, secrete their contents into ducts that may be lined with striated epithelial cells indicative of ion exchange activity. (×100)

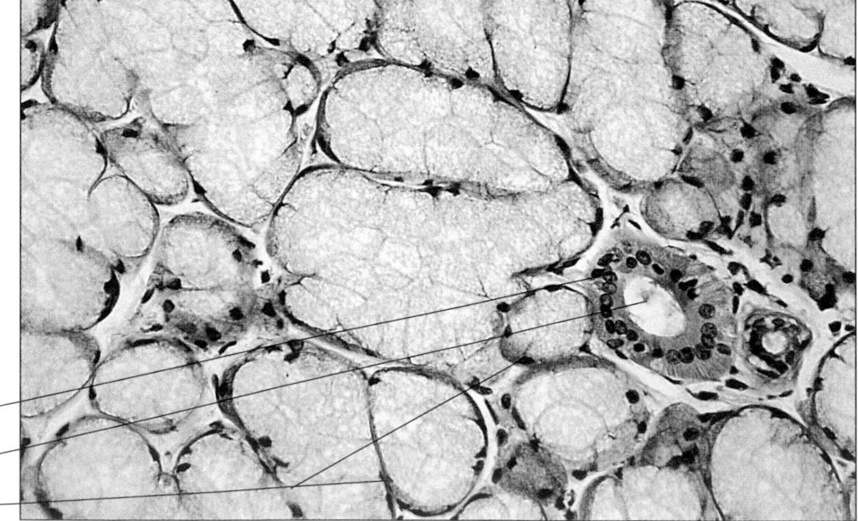

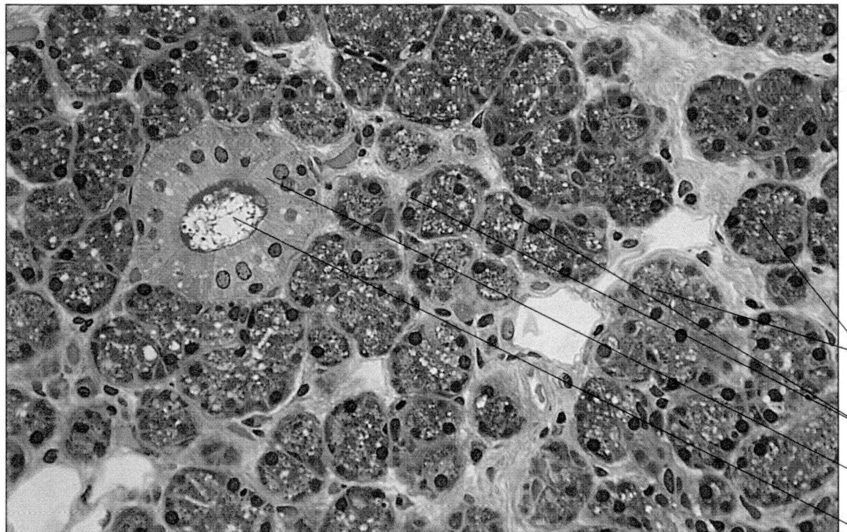

Figure 1-88
Parotid Salivary Gland Granular serous cells with numerous, large, zymogen granules surround duct. Several tiny ducts run between clusters within the plane of section. Human. (×100)

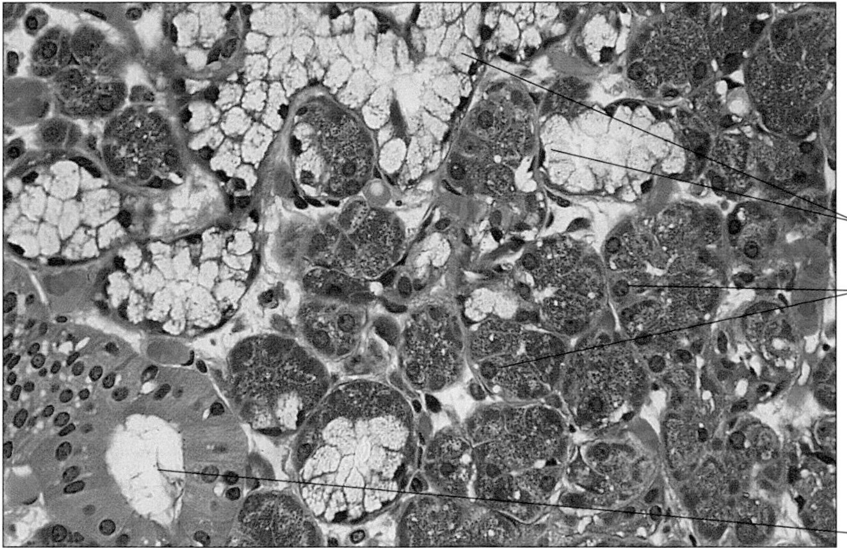

Figure 1-89
Submandibular Salivary Gland with Mucous (Light Staining) and Serous (Dark Staining) Components Striated duct is visible at lower left. (×100)

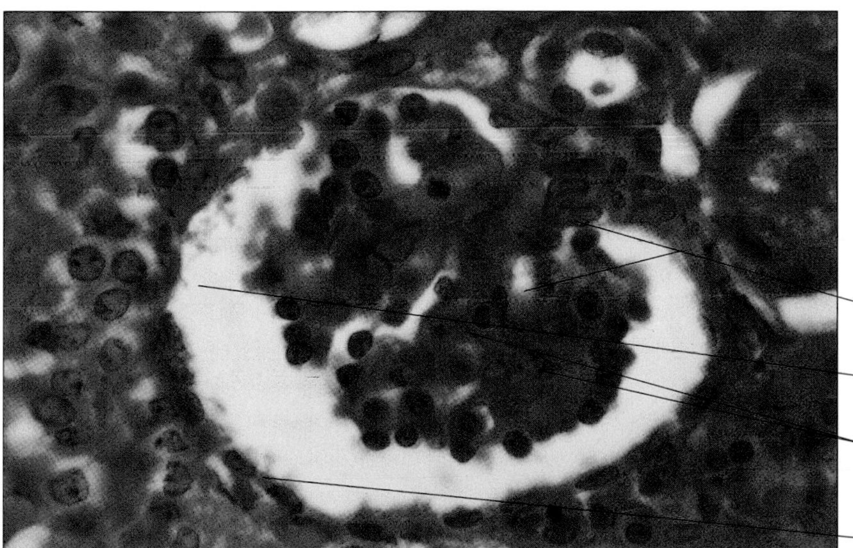

Figure 1-90
Glomerular (Bowman's) Capsule and Glomerulus (Renal Corpuscle) Tuft of capillaries, surrounded by podocytes, protrudes into space of glomerular capsule. Parietal surface is lined with single layer of simple squamous cells. (×100)

Figure 1-91
Two Glomeruli and Glomerular Capsules
"Lacy" edges of glomerulus on left shows characteristics of pregnancy-induced hypertension (PIH), here induced experimentally in a pregnant rat. (×50)

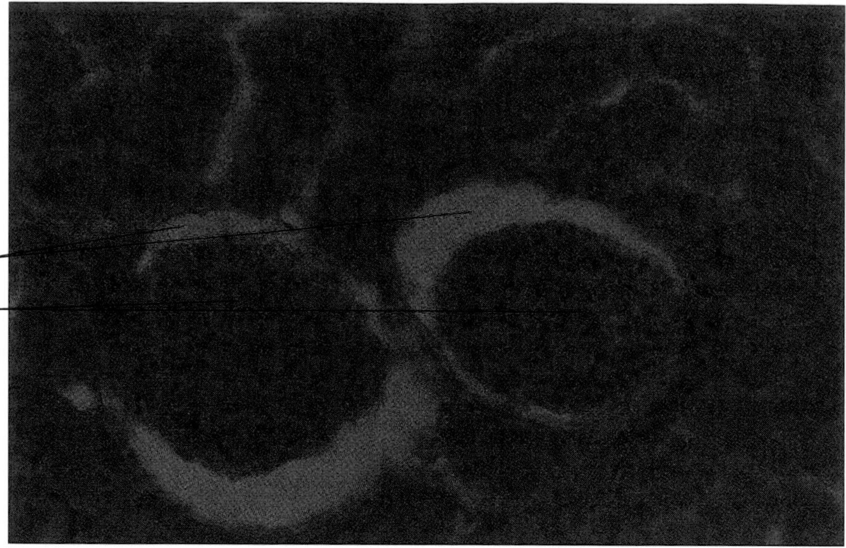

Figure 1-92
Distal Convoluted Tubules Lined with Cuboidal Epithelium Cross section from rat. (×400)

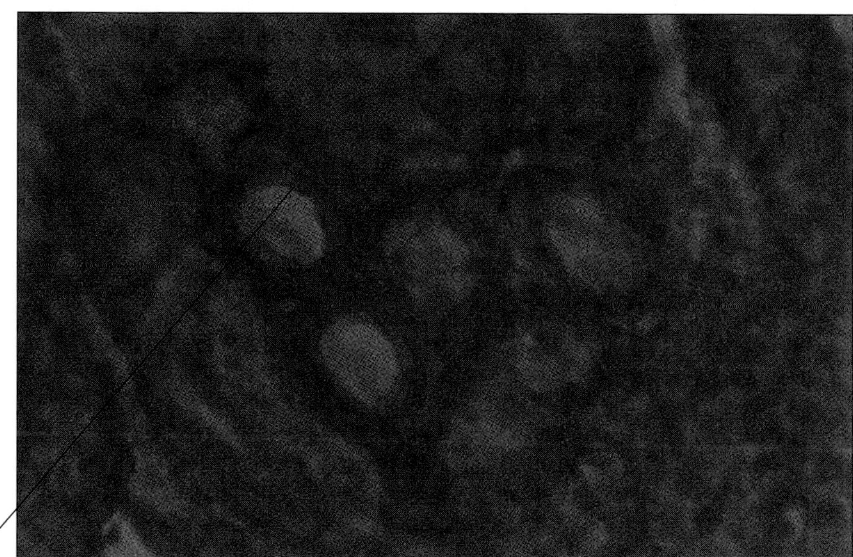

Figure 1-93
Ureter Star-shaped lumen is lined with transitional epithelium that varies in thickness to change shape as lumen stretches. Delicate lamina propria separates epithelium from alternating layers of circular and longitudinal smooth muscle. (×25)

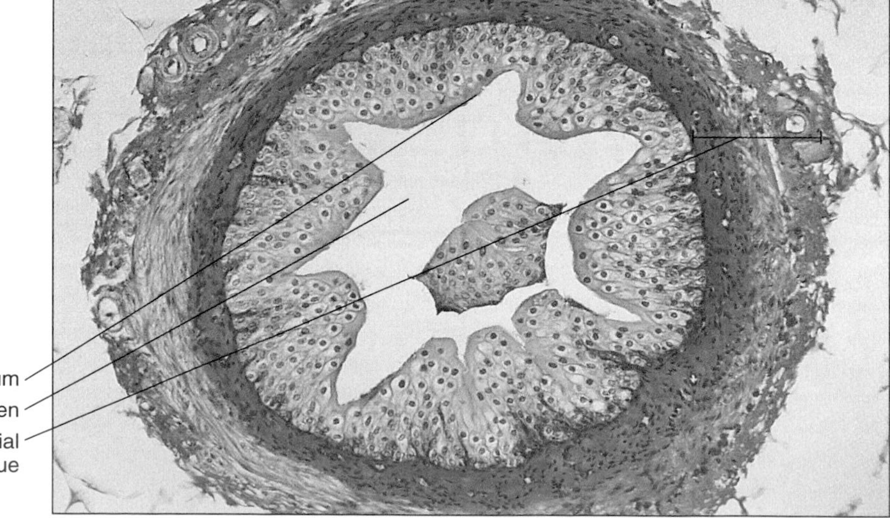

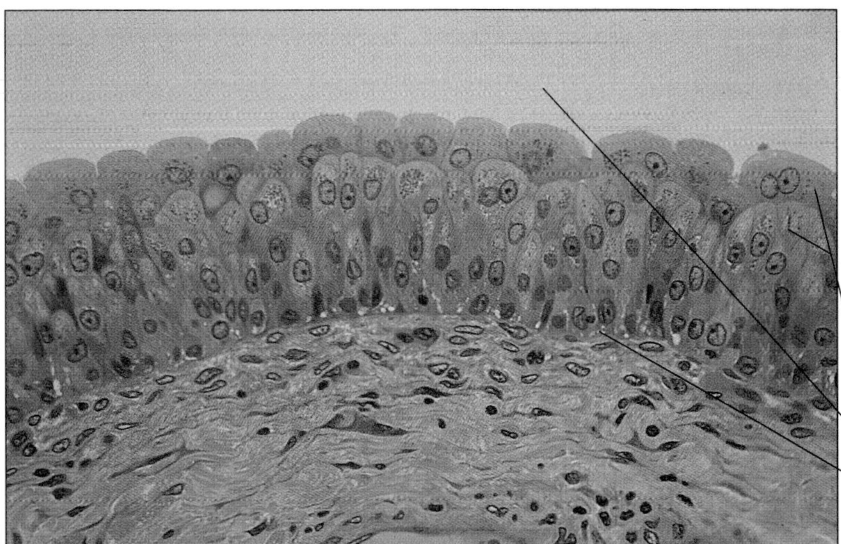

Figure 1-94

Urinary Bladder Umbrella cells of transitional epithelium stretch and flatten as bladder fills. Basement membrane separates epithelium from underlying connective tissue containing blood vessels. Monkey. (×100)

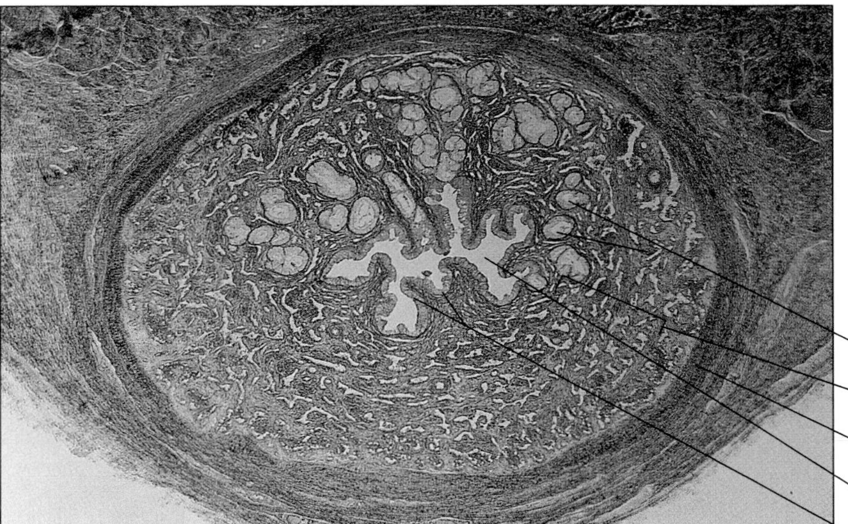

Figure 1-95

Urethra (within Penis) Lumen is lined with transitional epithelium and is embedded in corpus spongiosum of the penis. Paraurethral glands located above the lumen in the figure secrete mucus into the urethra. A smooth muscle layer (tunica muscularis) surrounds the urethral structures. (×10)

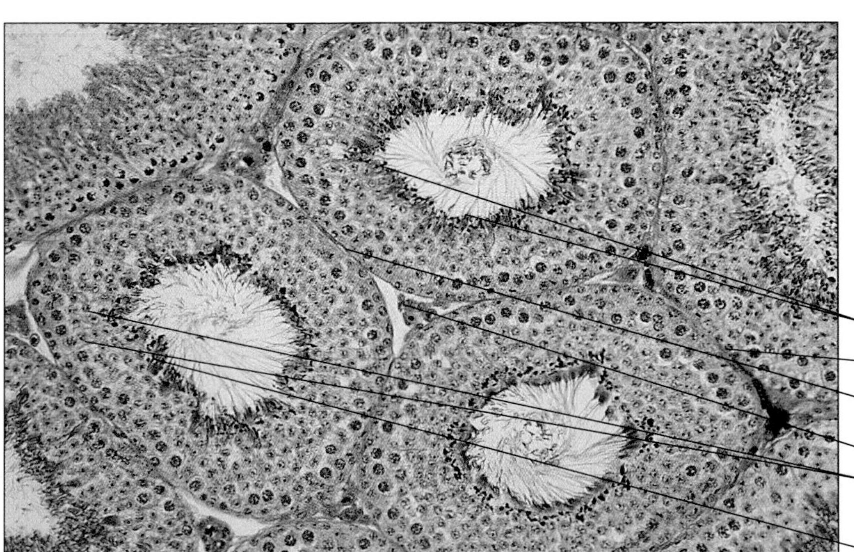

Figure 1-96

Seminiferous Tubules of Testis Lined with Sertoli Cells and Germinativum in Various Stages of Development Tunica propria surrounds each tubule. Interstitial spaces contain blood vessels and clumps of interstitial (Leydig) cells that secrete testosterone. (×50)

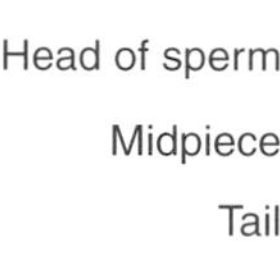

Figure 1-97
Spermatozoa Head contains numerous enzymes and nucleus with DNA. Thick midpiece just behind head is packed with mitochondria. (×250)

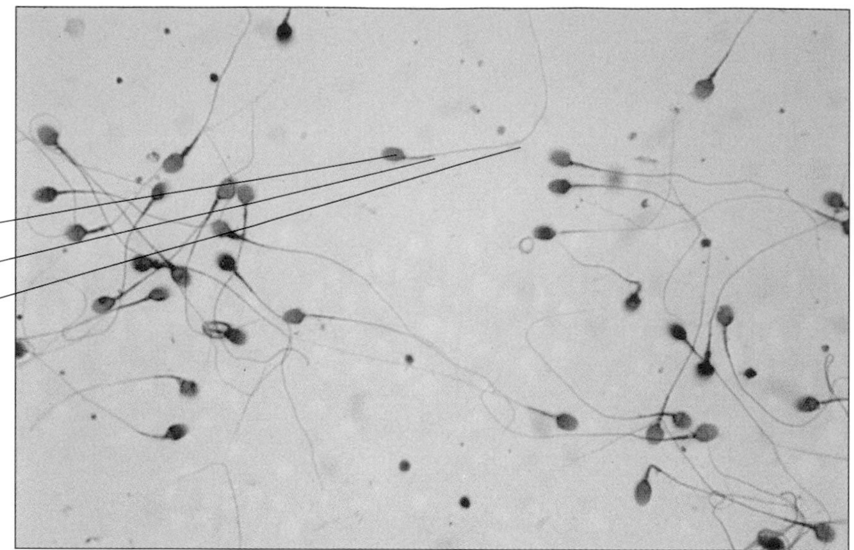

Figure 1-98
Epididymis Tall, pseudostratified columnar epithelium with microvilli surrounds a lumen packed with clumps of spermatozoa. Narrow band of smooth muscle cells encircles each tubule.

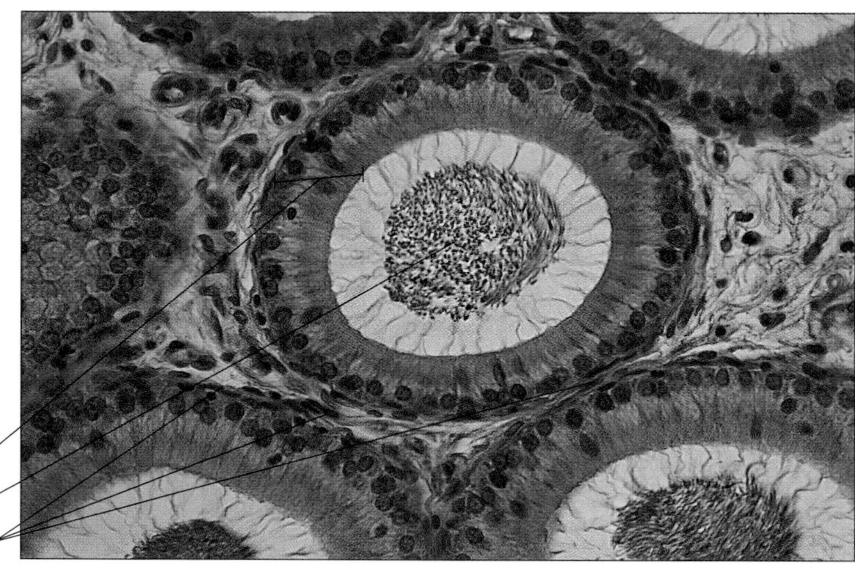

Figure 1-99
Ductus Deferens Ciliated columnar epithelial cells line a spermatozoa-filled lumen. Three layers of smooth muscle cells surround mucosa, a circular layer between two longitudinal ones. (×50)

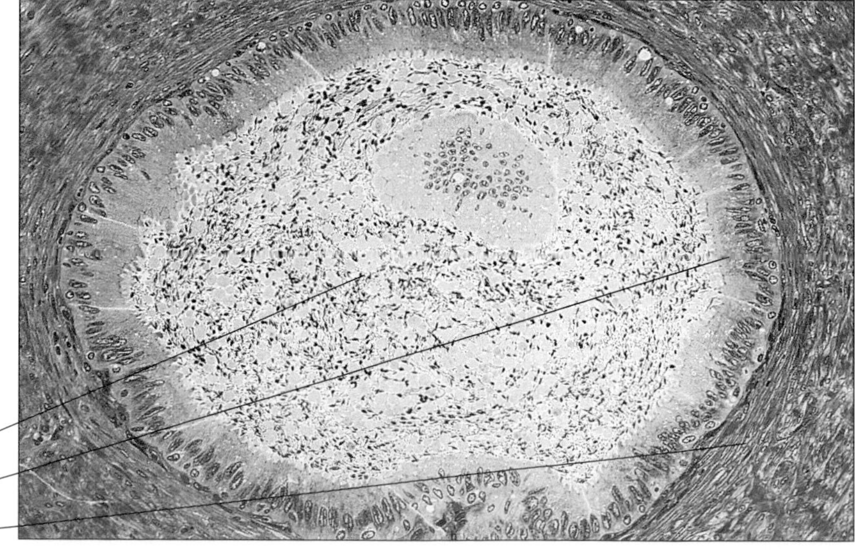

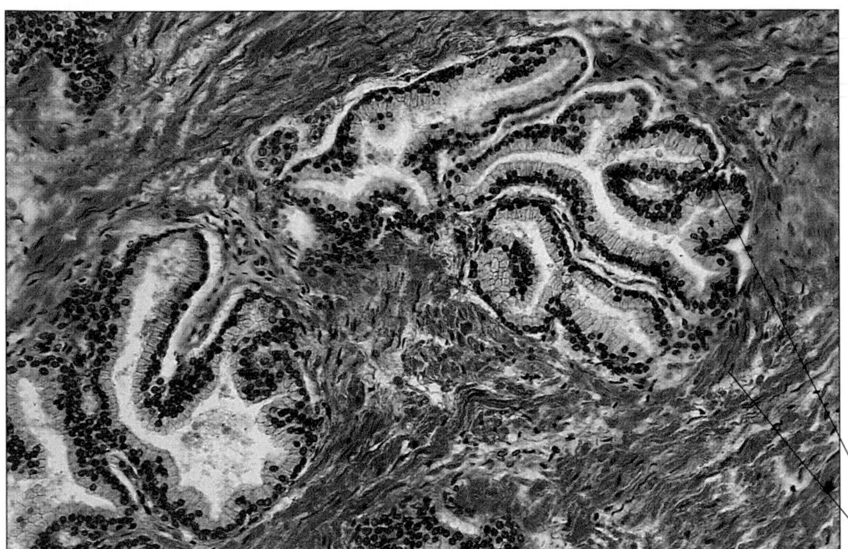

Figure 1-100
Prostate Gland Mucosal surfaces, lined with tall columnar cells and darkly stained basal nuclei, are arranged in numerous deep folds. Lumina open directly into prostatic urethra. Smooth muscle and fibrocollagenous stroma surround luminal structures. Human. (×50)

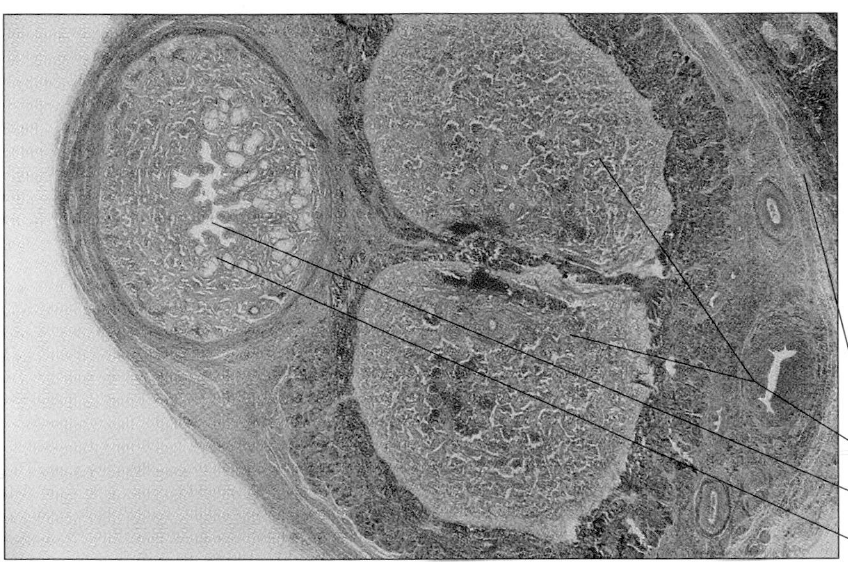

Figure 1-101
Penis Two corpora cavernosa lie superior to single corpus spongiosum containing penile urethra. Septum between corpora cavernosa is incomplete. Dense fibrous connective tissue, tunica albuginea, surrounds the three vascular cavernosa. The inferior aspect appears on the left, the superior aspect on the right. (×5)

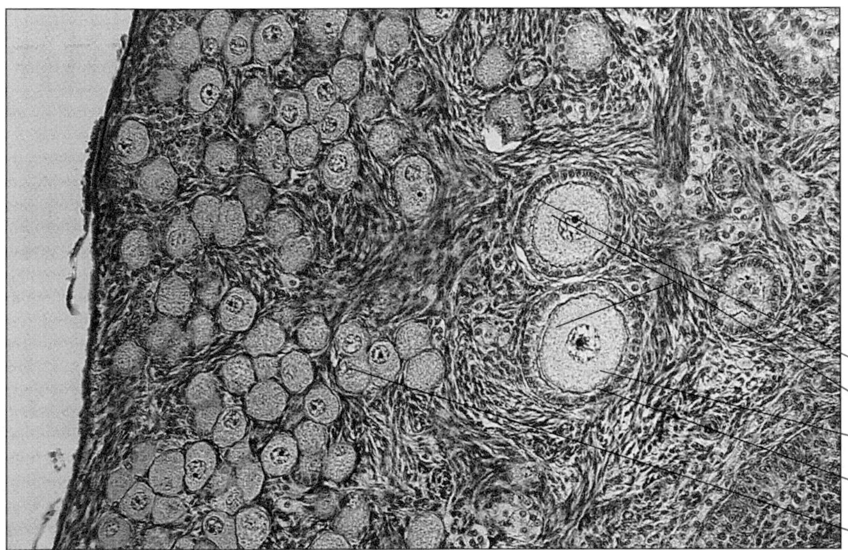

Figure 1-102
Ovary with Numerous Primordial Follicles and Two Primary Follicles Primordial follicles contain oocytes that are not stimulated to complete the first meiotic division. Two primary follicles each contain an ovum with nucleus and clear surrounding cytoplasm. Thin, clear **zona pellucida** is surrounded by a ring of even cuboidal cells, the **corona radiata.** (×25)

Figure 1-103
Detail of Oocyte in Primordial Follicle
Clear nucleus contains well-defined nucleolus. Neither zona pellucida nor corona radiata is evident. (×250)

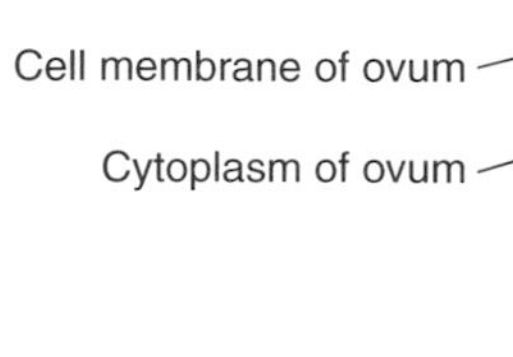

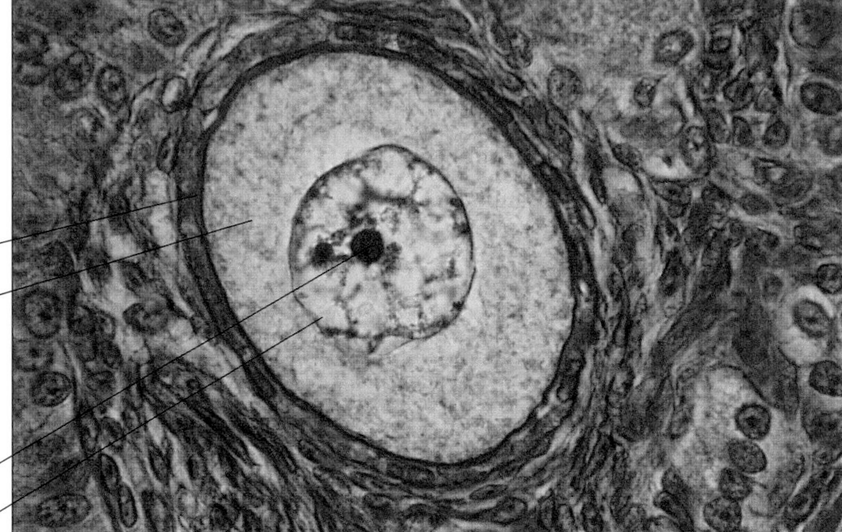

Figure 1-104
Secondary Ovarian Follicle with Ovum
Bright zona pellucida surrounds outer membrane of ovum and in turn is surrounded by dark, cellular corona radiata. A large **antrum** has formed where the egg is not anchored to the follicular wall of **granulosa cells.** (×100)

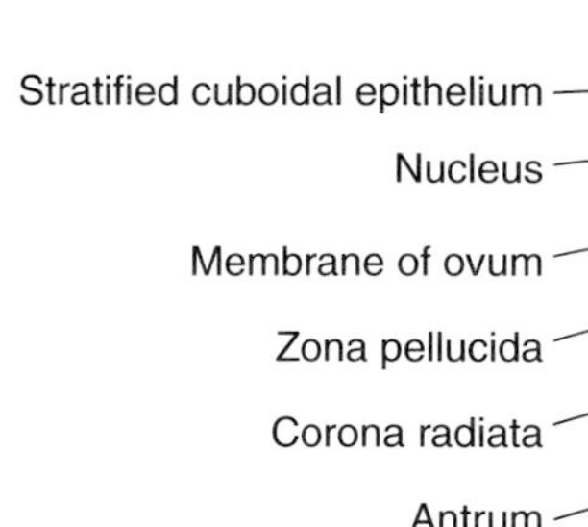

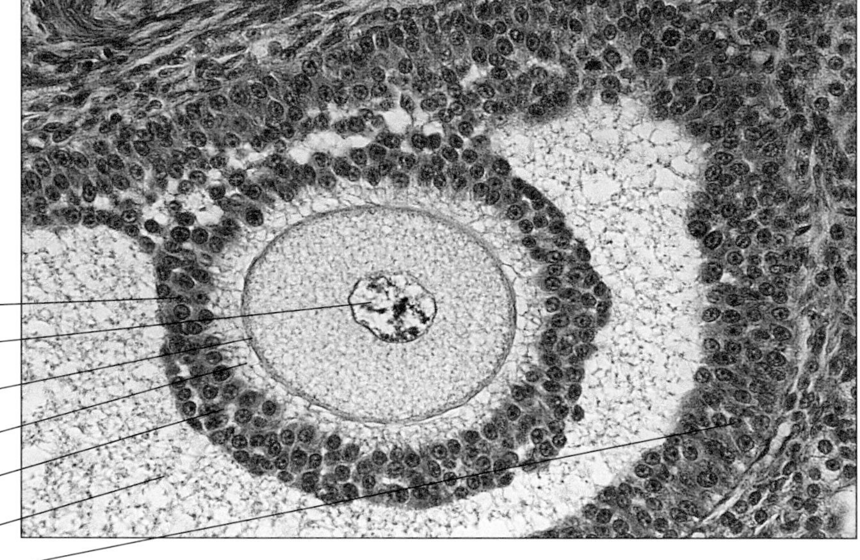

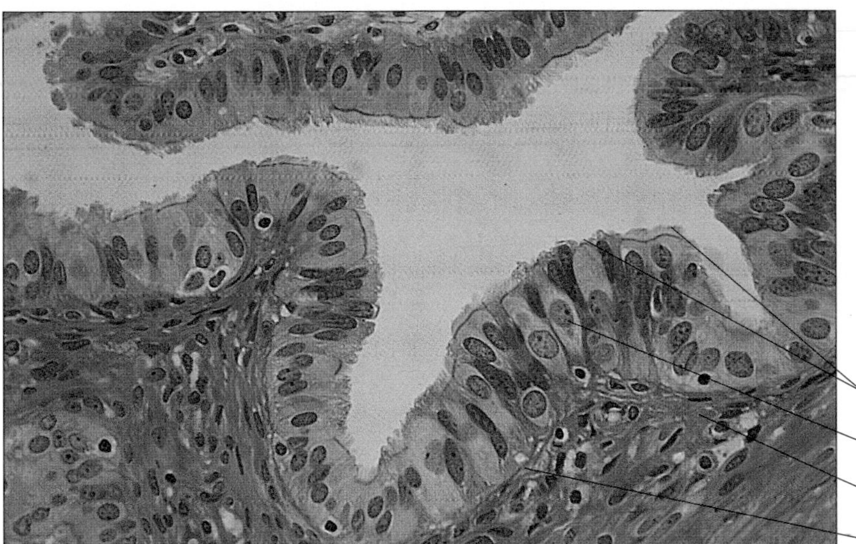

Figure 1-105
Fallopian (Uterine) Tube Extensive folding of mucosa, lined with ciliated columnar epithelium, is common. Epithelium rests on thin basement membrane and flat connective tissue layer. Rhythmic beating of cilia helps transport ovum toward uterus; cell structure also suggests secretory function. Human. (×100)

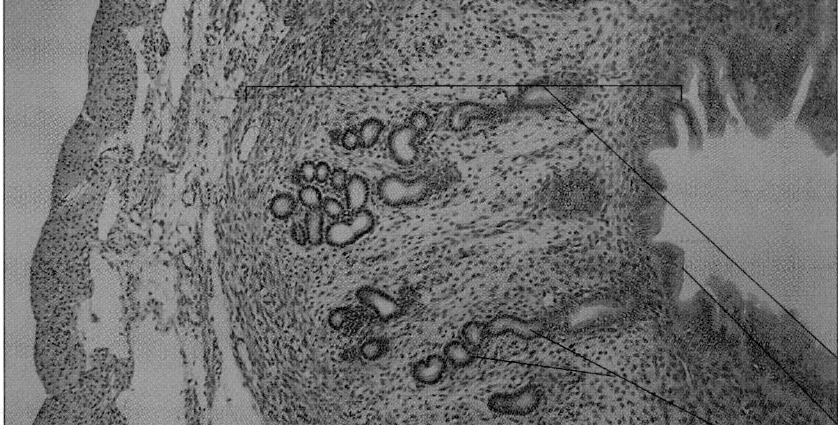

Figure 1-106
Uterus Endometrial lining *(right)* during proliferative phase of uterine cycle shows thickening of epithelial surfaces and numerous coiled glandular ducts. (×25)

CHAPTER 2

Human Skeletal Anatomy

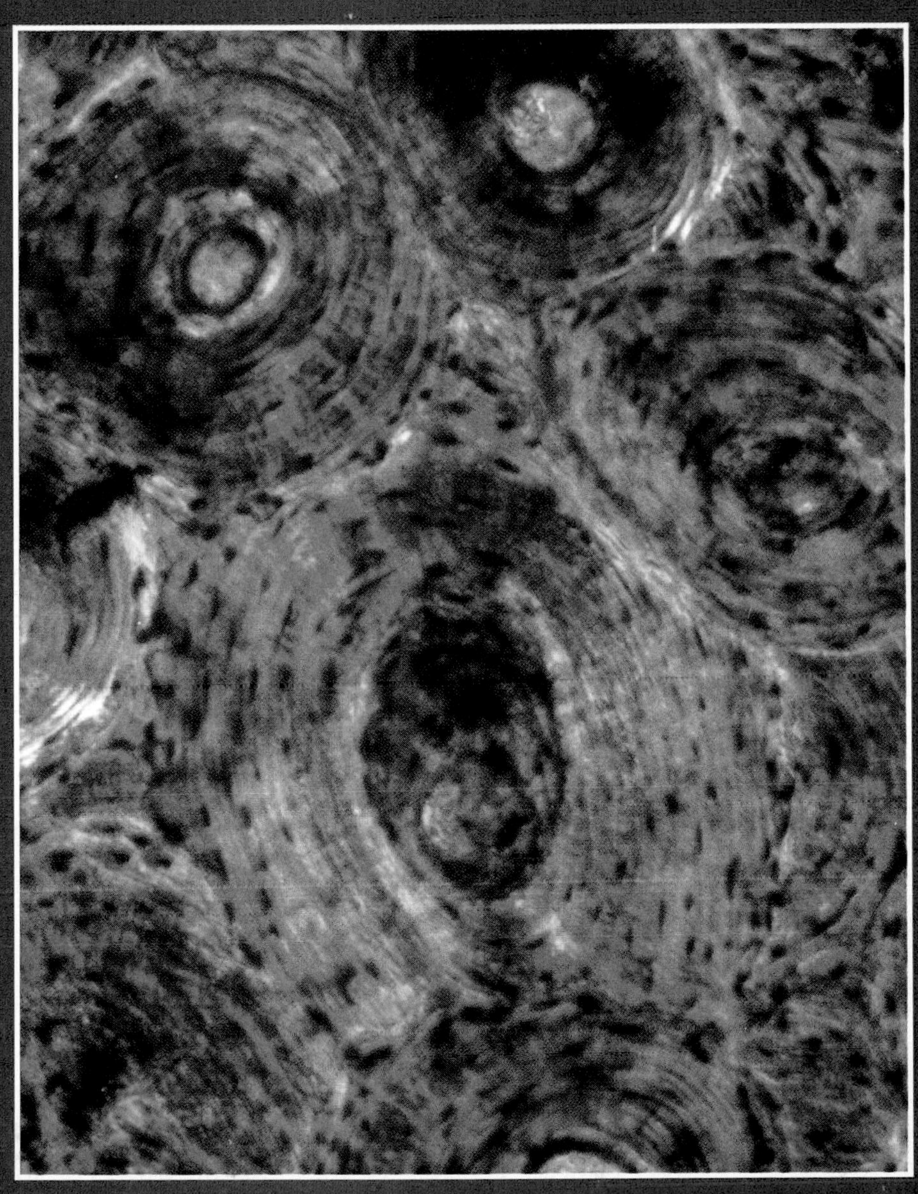

Compact Bone.
Light micrograph, magnification: ×265

Figure 2-1
Skull: Anterior View

BONES

1. Frontal
2. Inferior nasal concha
3. Lacrimal
4. Mandible
5. Maxilla
6. Nasal
7. Parietal
8. Sphenoid
9. Temporal
10. Vomer
11. Zygomatic

FORAMINA

12. Inferior orbital fissure
13. Infra-orbital foramen of maxilla
14. Lacrimal canal
15. Mental foramen of mandible
16. Optic canal of sphenoid
17. Superior orbital fissure
18. Supra-orbital foramen of frontal bone

PROCESSES

19. Anterior nasal spine of maxilla
20. Mandibular alveolus
21. Maxillary alveolus
22. Perpendicular plate of ethmoid
23. Ramus of mandible
24. Supra-orbital notch of frontal bone
25. Zygomatic process of maxilla

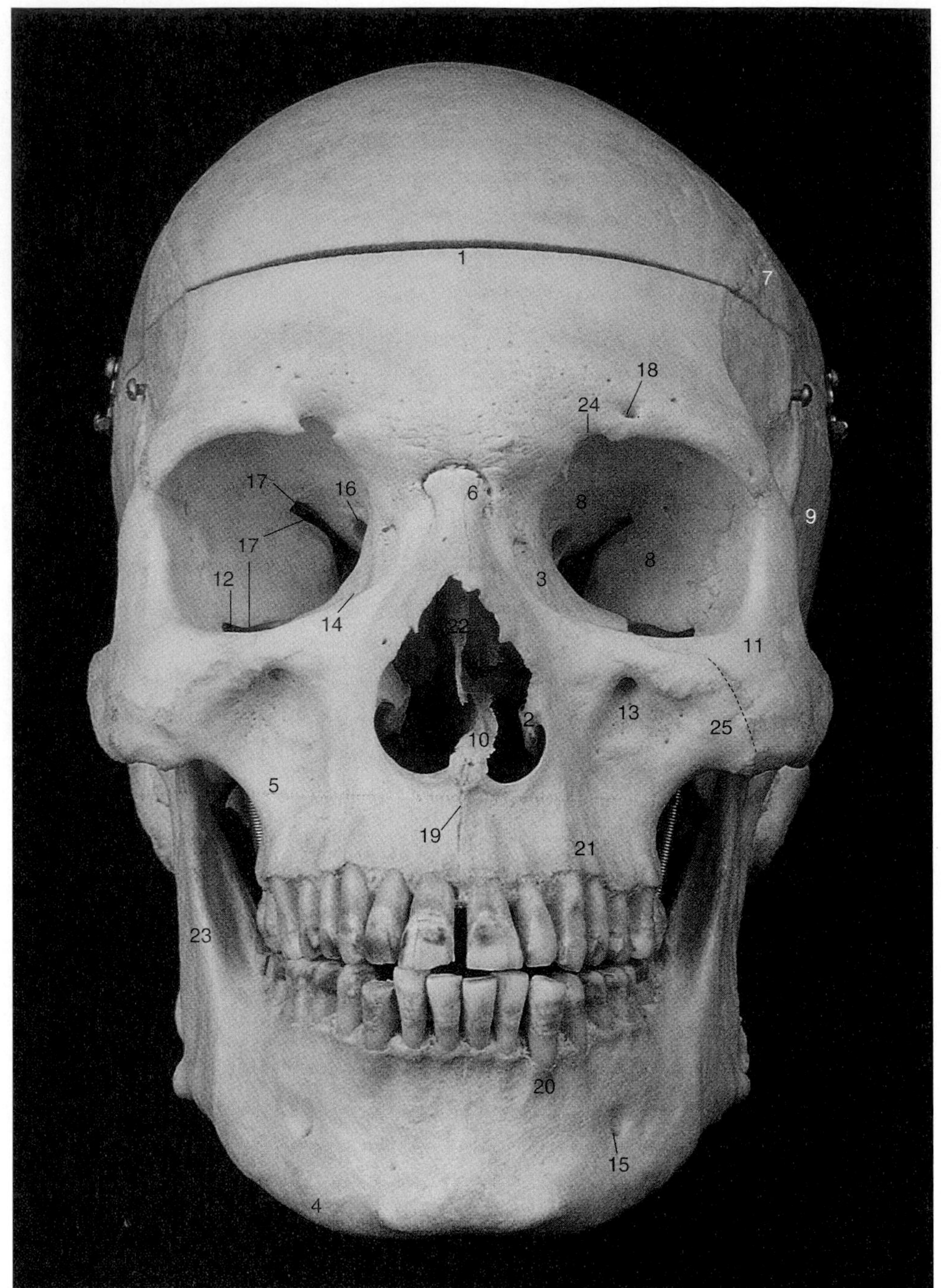

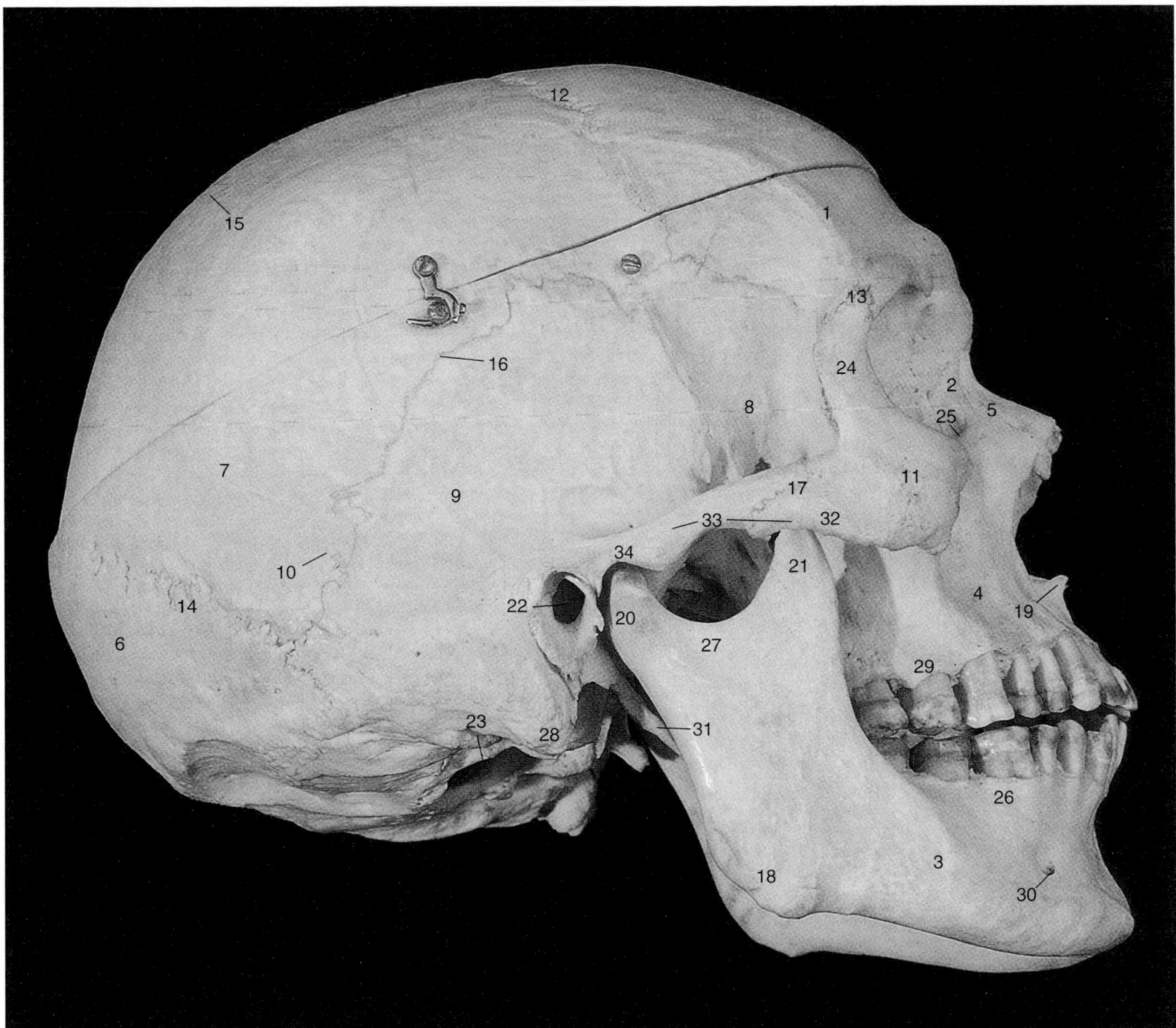

Figure 2-2
Skull: Lateral View

BONES

1. Frontal
2. Lacrimal
3. Mandible
4. Maxilla
5. Nasal
6. Occipital
7. Parietal
8. Sphenoid (greater wing)
9. Temporal
10. Wormian (small bones between sutures)
11. Zygomatic

SUTURES

12. Coronal
13. Frontozygomatic
14. Lambdoid
15. Sagittal
16. Squamous
17. Temporozygomatic

FORAMINA & PROCESSES

18. Angle of mandible
19. Anterior nasal spine of maxilla
20. Condylar process of mandible
21. Coronoid process of mandible
22. External acoustic meatus of temporal bone
23. Foramen magnum of occipital bone
24. Frontal process of zygomatic bone
25. Lacrimal groove
26. Mandibular alveolus of mandible
27. Mandibular notch
28. Mastoid process of temporal bone
29. Maxillary alveolus of maxilla
30. Mental foramen of mandible
31. Styloid process of temporal bone
32. Temporal process of zygomatic bone
33. Zygomatic arch
34. Zygomatic process of temporal bone

Figure 2-3
Skull: Calvarium, Superior View

BONES
1. Frontal
2. Occipital
3. Parietal

SUTURES
4. Bregma
5. Coronal
6. Sagittal

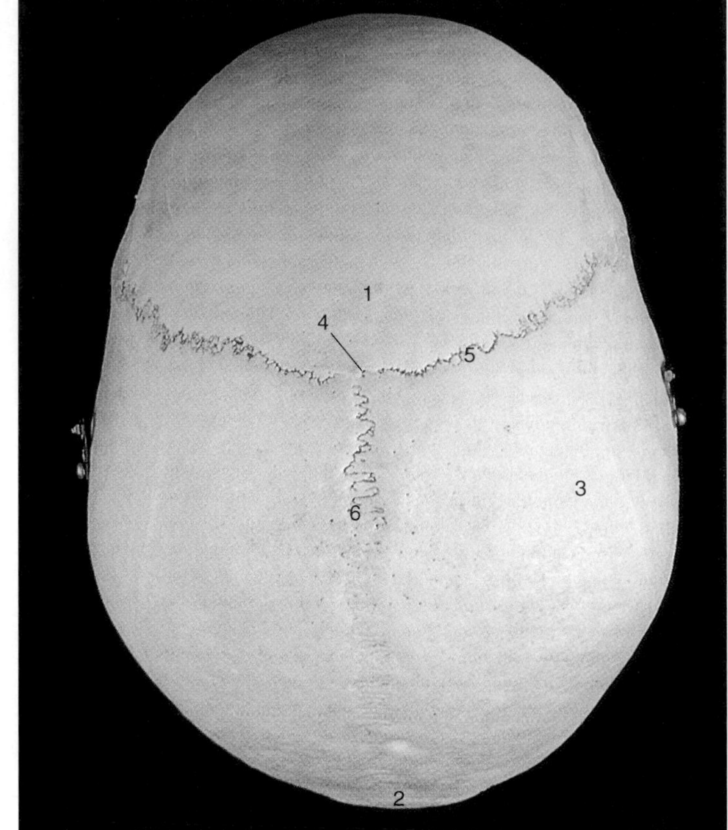

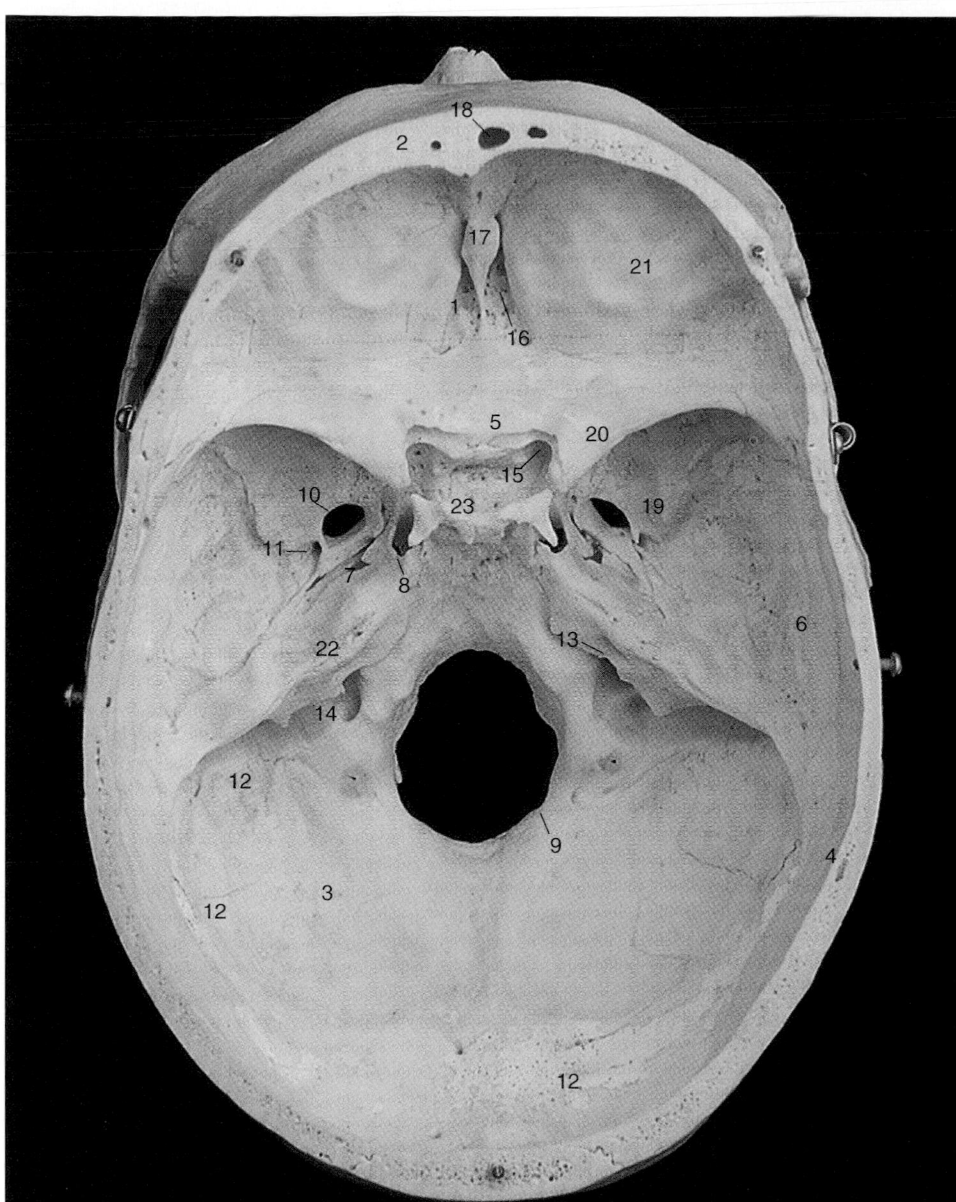

Figure 2-4
Skull: Floor of Cranium, Internal View

BONES

1. Ethmoid
2. Frontal
3. Occipital
4. Parietal
5. Sphenoid
6. Temporal

FORAMINA & DEPRESSIONS

7. Carotid canal of temporal bone
8. Foramen lacerum
9. Foramen magnum of occipital bone
10. Foramen ovale of sphenoid
11. Foramen spinosum of sphenoid
12. Grooves for transverse and sigmoid sinuses
13. Internal acoustic meatus of temporal bone
14. Jugular foramen of temporal bone
15. Optic canal of sphenoid

PROCESSES

16. Cribriform plate of ethmoid
17. Crista galli of ethmoid
18. Frontal sinus
19. Greater wing of sphenoid
20. Lesser wing of sphenoid
21. Orbital plate of frontal bone
22. Petrous portion of temporal bone (The incus, malleus, and stapes are not shown. They reside within the petrous portion of the temporal bone.)
23. Sella turcica of sphenoid

Figure 2-5
Skull: Base, Inferior View

BONES

1. Maxilla (palatine process)
2. Occipital
3. Palatine
4. Temporal
5. Vomer
6. Zygomatic

FORAMINA

7. Carotid canal of temporal bone
8. Condylar fossa and canal of occipital bone
9. Foramen lacerum
10. Foramen magnum of occipital bone
11. Foramen ovale of sphenoid
12. Foramen spinosum of sphenoid
13. Greater palatine foramen
14. Hypoglossal canal of occipital bone
15. Incisive canal of maxilla
16. Inferior orbital fissure
17. Jugular fossa of temporal bone
18. Stylomastoid foramen of temporal bone

PROCESSES, DEPRESSIONS, & SUTURES

19. External occipital crest
20. External occipital protruberance of occipital bone
21. Highest nuchal line of occipital bone
22. Inferior nuchal line of occipital bone
23. Lesser palatine foramina of palatine bone
24. Mandibular fossa of temporal bone
25. Mastoid process of temporal bone
26. Medial and lateral pterygoid processes of sphenoid
27. Occipital condyle
28. Occipitotemporal suture
29. Posterior nasal spine of palatine bone
30. Spheno-occipital synchondrosis
31. Styloid process of temporal bone
32. Superior nuchal line of occipital bone
33. Temporal process of zygomatic bone
34. Zygomatic arch
35. Zygomatic process of temporal bone

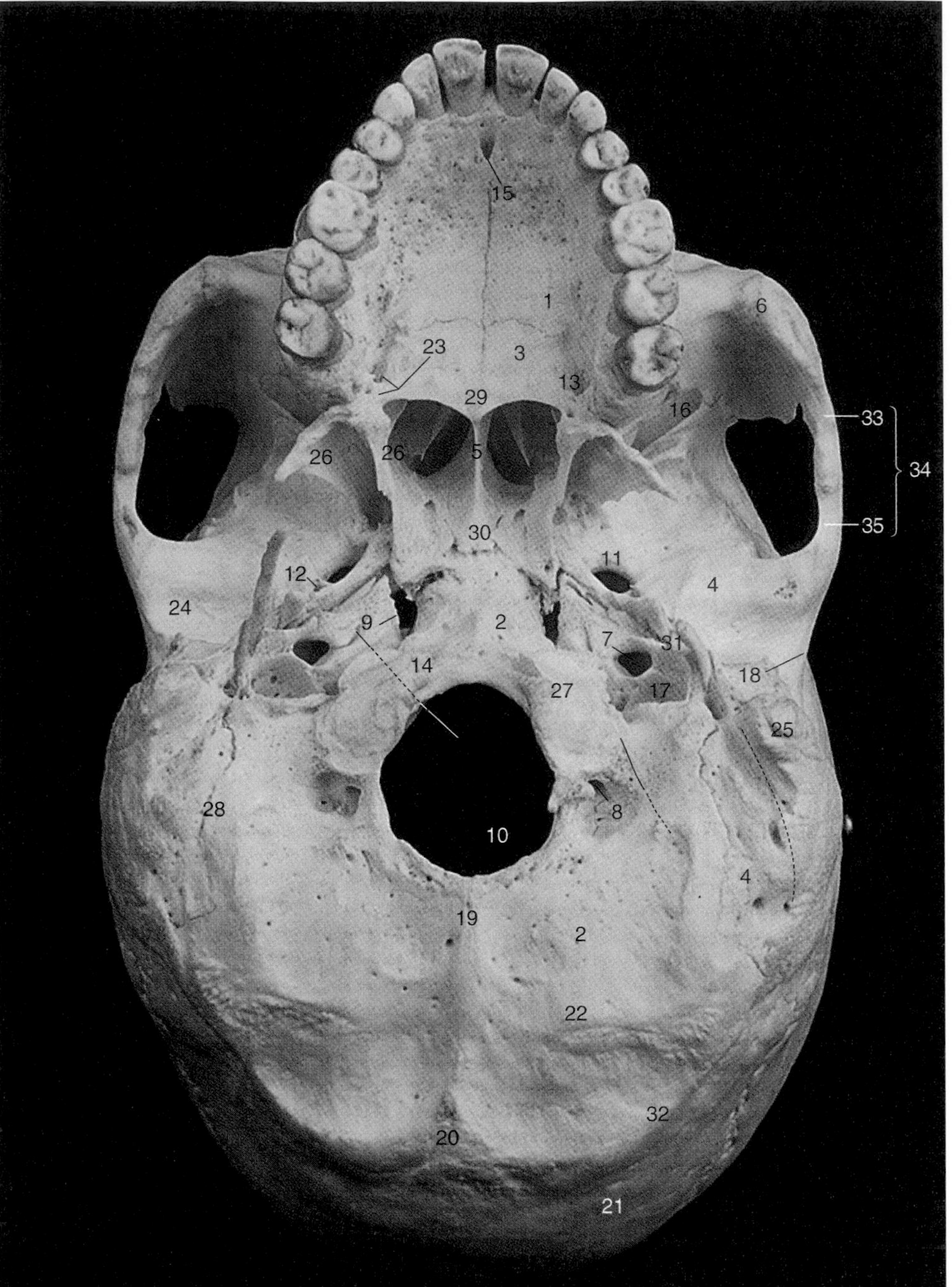

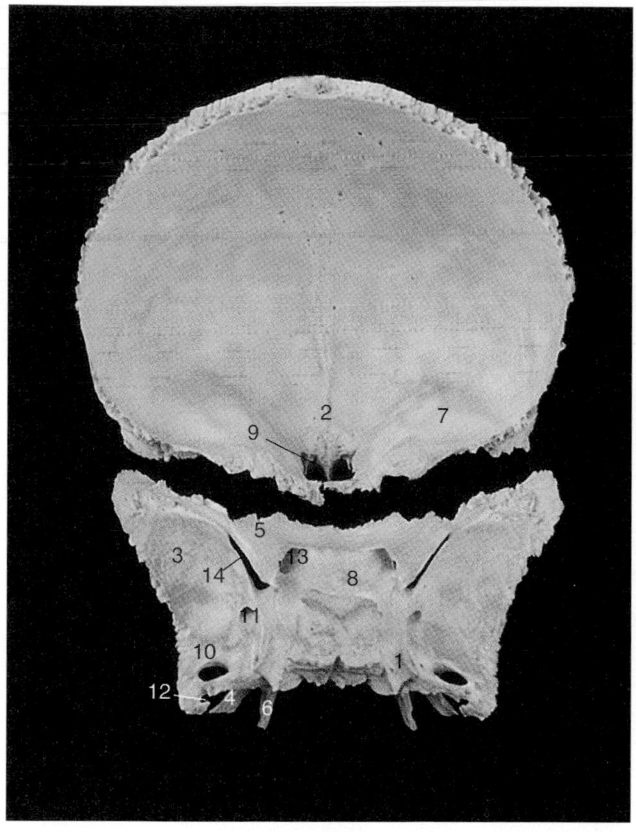

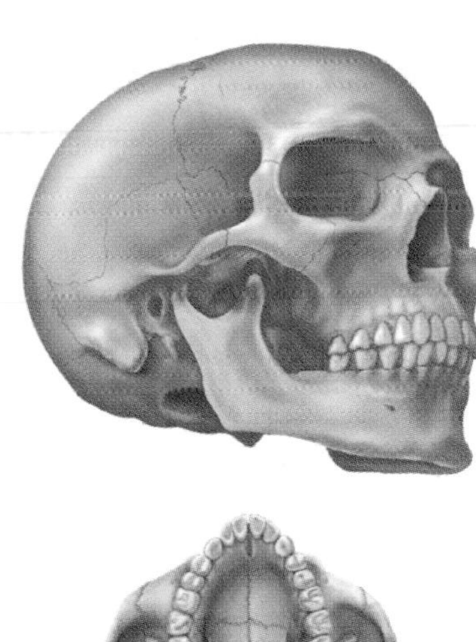

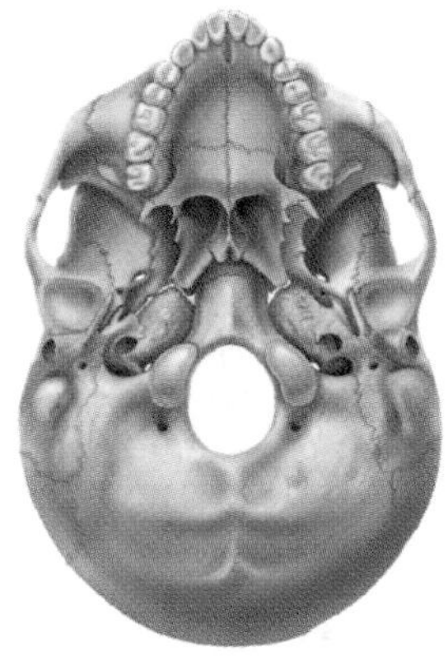

(external views)

Figure 2-6
Frontal and Sphenoid Bones: Internal View, Frontosphenoidal Suture Separated

PROCESSES & DEPRESSIONS

1. Carotid sulcus of sphenoid
2. Frontal crest of frontal bone
3. Greater wing of sphenoid
4. Lateral plate of pterygoid process of sphenoid
5. Lesser wing of sphenoid
6. Medial plate of pterygoid process of sphenoid
7. Orbital part of frontal bone
8. Sella turcica of sphenoid

FORAMINA

9. Ethmoidal notch of frontal bone
10. Foramen ovale of sphenoid
11. Foramen rotundum of sphenoid
12. Foramen spinosum of sphenoid
13. Optic canal of sphenoid
14. Superior orbital fissure of sphenoid

Figure 2-7
Sphenoid: Anterior View

PROCESSES

1. Lateral plate of pterygoid process of sphenoid
2. Medial plate of pterygoid process of sphenoid
3. Orbital surface of sphenoid
4. Rostrum of sphenoid
5. Temporal surface of sphenoid

FORAMINA

6. Foramen ovale of sphenoid
7. Foramen rotundum of sphenoid
8. Foramen spinosum of sphenoid
9. Optic canal of sphenoid
10. Pterygoid canal of sphenoid
11. Superior orbital fissure of sphenoid

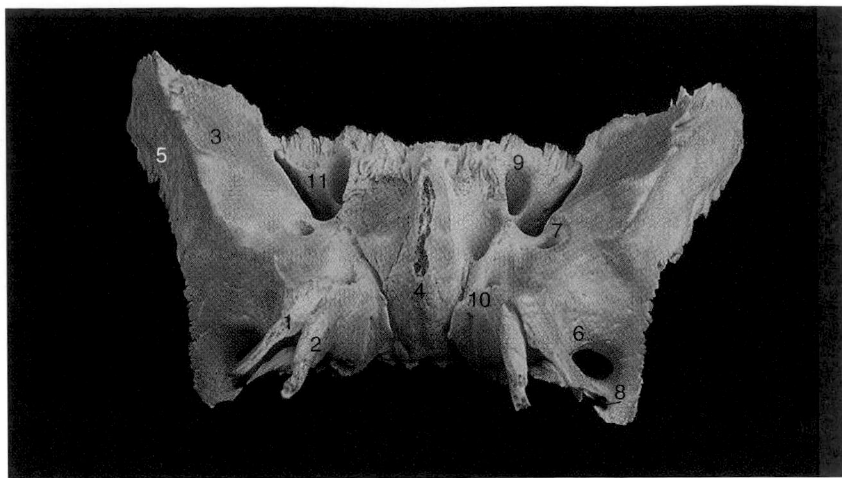

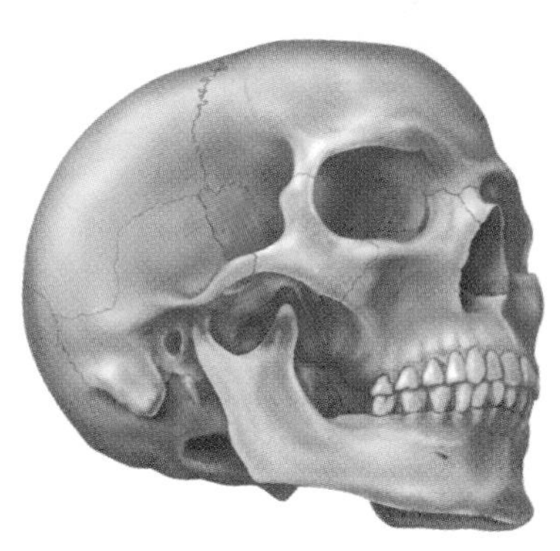

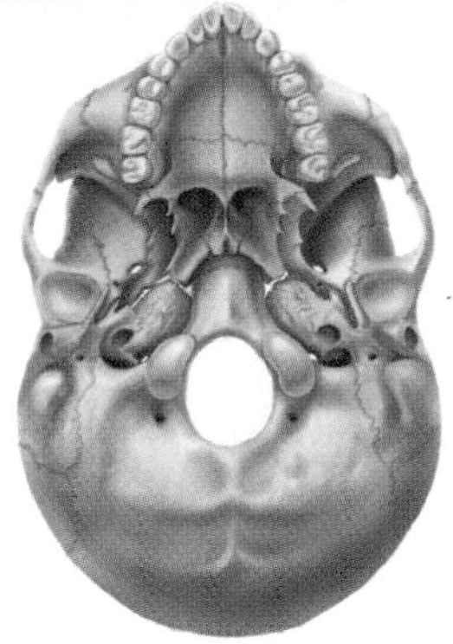

Figure 2-8
Frontal and Zygomatic Bones: Anterior View

PROCESSES

1. Frontal process of zygomatic bone
2. Maxillary border of zygomatic bone
3. Orbital border of zygomatic bone
4. Orbital part of frontal bone
5. Temporal border of zygomatic bone
6. Temporal process of zygomatic bone
7. Zygomatic process of frontal bone

FORAMINA

8. Ethmoidal notch of frontal bone
9. Supra-orbital foramen of frontal bone
10. Supra-orbital notch of frontal bone
11. Zygomatico-orbital foramen of zygomatic bone

Note: Supra-orbital foramen and notch on right side are separate; on left they are superimposed.

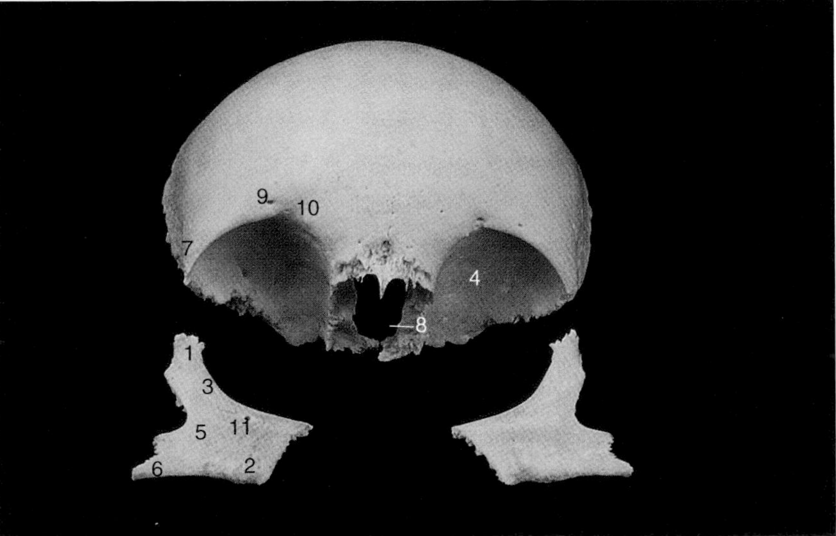

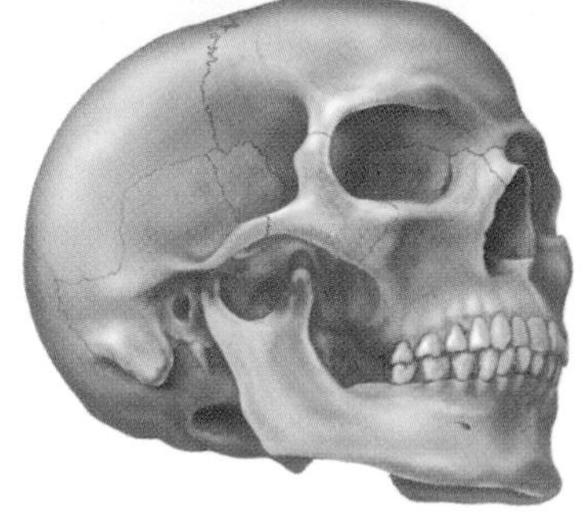

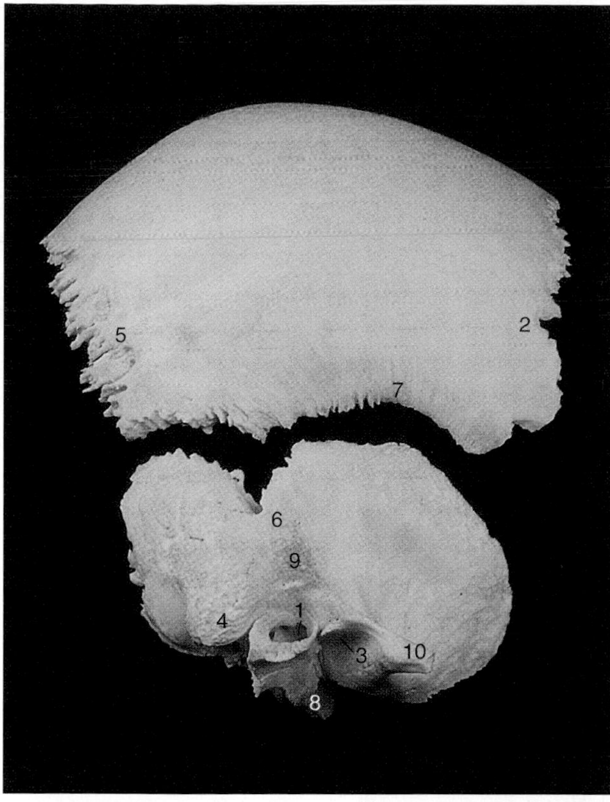

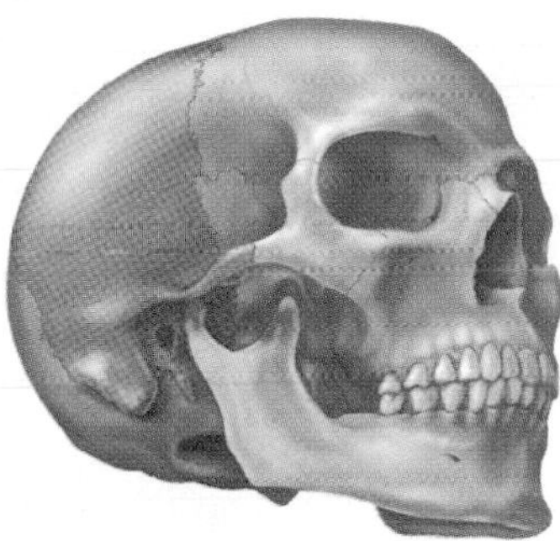

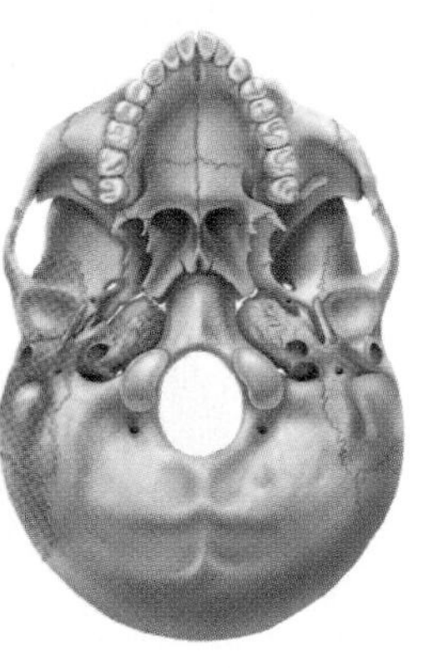

Figure 2-9
Right Temporal and Parietal Bones: Lateral View, Squamosal Suture Separated

STRUCTURES

1. External acoustic meatus of temporal bone
2. Frontal border of parietal bone
3. Mandibular fossa of temporal bone
4. Mastoid process of temporal bone
5. Occipital border of parietal bone
6. Parietal notch of temporal bone
7. Squamosal border of parietal bone
8. Styloid process of temporal bone
9. Suprameatal triangle of temporal bone
10. Zygomatic process of temporal bone

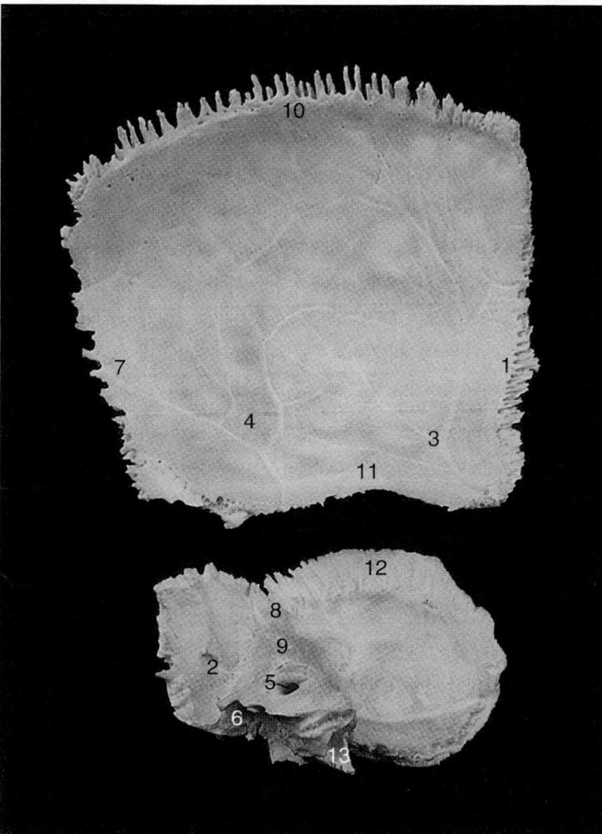

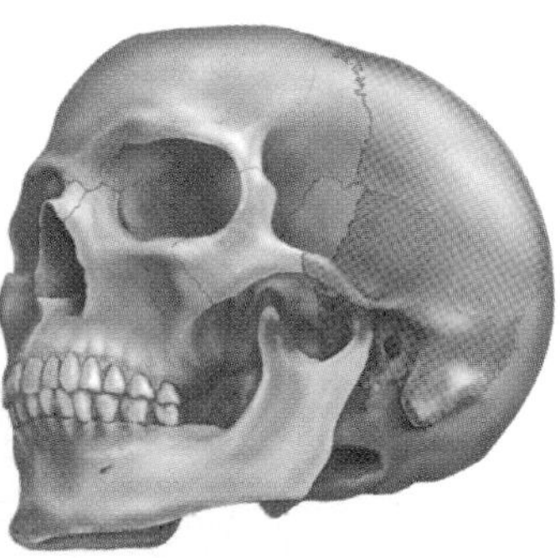

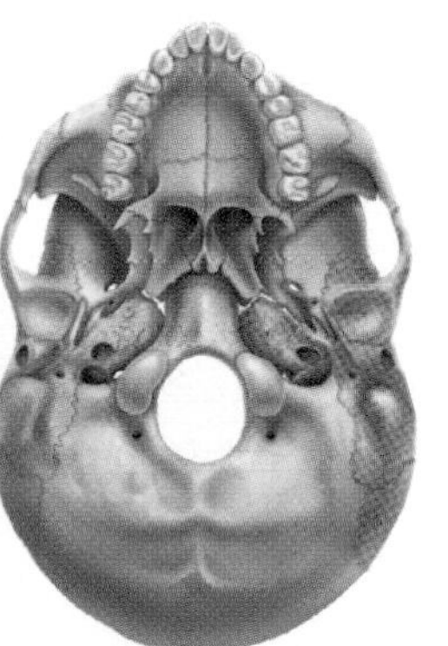

(external views)

Figure 2-10
Left Temporal and Parietal Bones: Internal View, Squamosal Suture Separated

STRUCTURES

1. Frontal border of parietal bone
2. Groove for sigmoid sinus of temporal bone
3. Grooves for frontal branch of middle meningeal vessels
4. Grooves for parietal branch of middle meningeal vessels
5. Internal acoustic meatus of temporal bone
6. Mastoid process of temporal bone
7. Occipital border of parietal bone
8. Parietal notch of temporal bone
9. Petrous portion of temporal bone
10. Sagittal border
11. Squamosal border of parietal bone
12. Squamosal border of temporal bone
13. Styloid process of temporal bone

Figure 2-11
Occipital Bone: Interior View

STRUCTURES

1. Cerebellar fossa
2. Cerebral fossa
3. Condylar fossa and canal
4. Foramen magnum
5. Groove for sigmoid sinus
6. Groove for superior sagittal sinus
7. Groove for transverse sinus
8. Internal occipital crest
9. Internal occipital protuberance
10. Jugular notch
11. Jugular tubercle
12. Lambdoidal margin
13. Mastoid margin

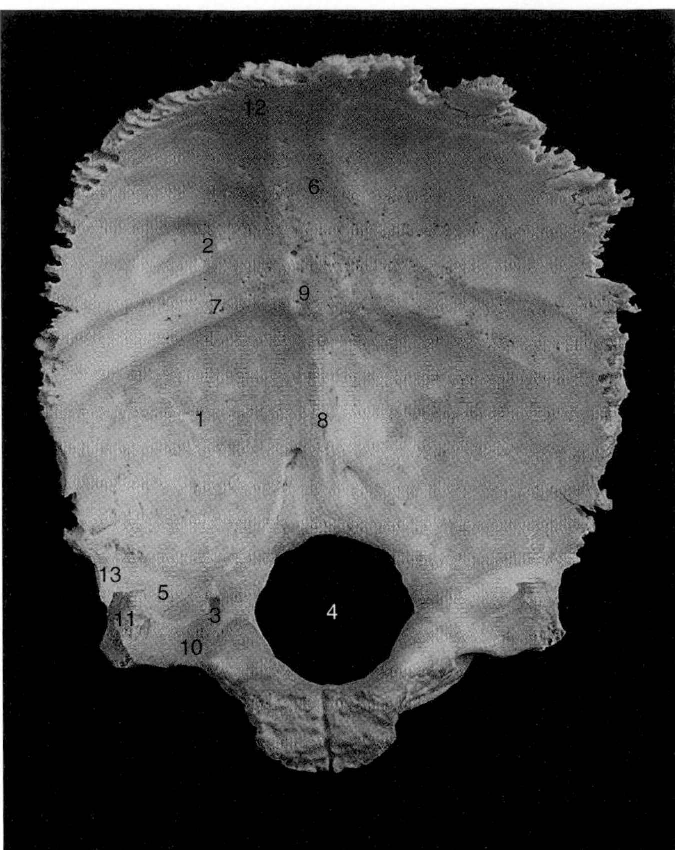

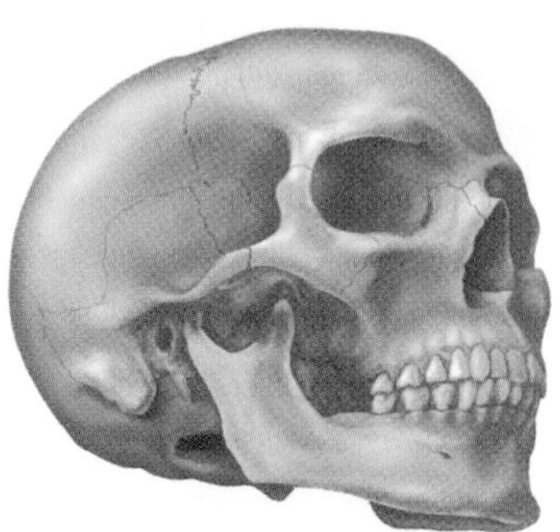

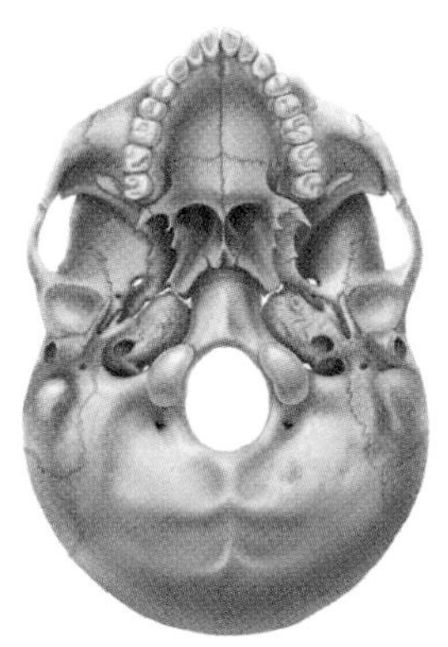

Figure 2-12
Palatine Bones: (A) Right from Anteriomedial View, (B) Left from Anteriosuperior View

STRUCTURES

A. Right palatine bone
B. Left palatine bone
1. Ethmoidal crest of palatine bone
2. Horizontal plate of palatine bone
3. Maxillary process of palatine bone
4. Orbital process of palatine bone
5. Perpendicular plate of palatine bone
6. Pyramidal process of palatine bone
7. Sphenoidal process of palatine bone
8. Sphenopalatine notch of palatine bone

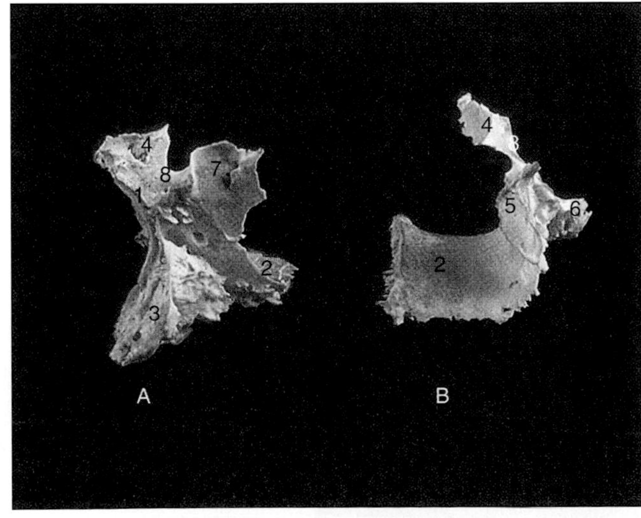

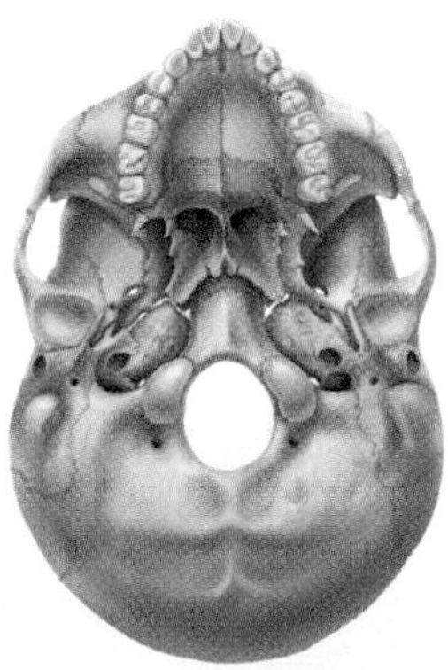

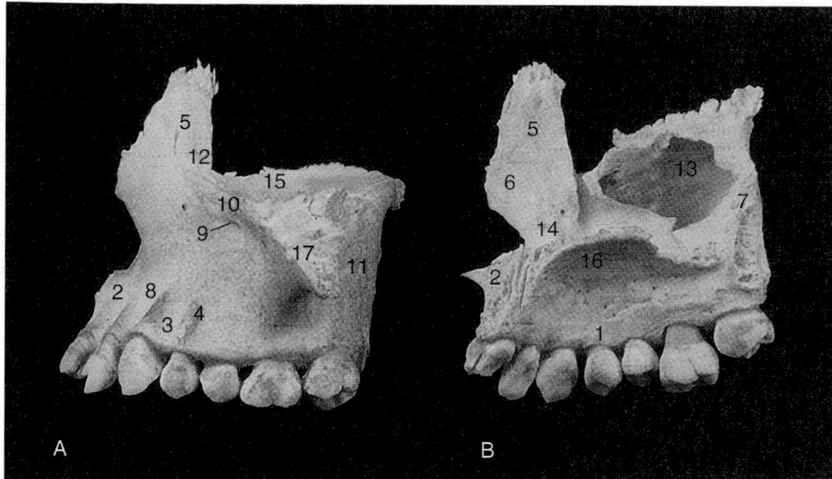

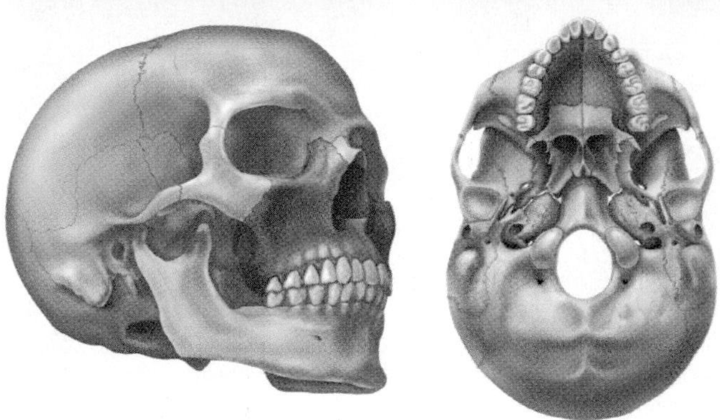

Figure 2-13
Maxillae: (A) Left from Lateral View, (B) Right from Medial View

STRUCTURES

A. Left maxilla
B. Right maxilla
1. Alveolar process of maxilla
2. Anterior nasal spine of maxilla
3. Canine eminence of maxilla
4. Canine fossa of maxilla
5. Ethmoidal crest of maxilla
6. Frontal process of maxilla
7. Greater palatine canal of maxilla
8. Incisive fossa of maxilla
9. Infra-orbital foramen of maxilla
10. Infra-orbital margin of maxilla
11. Infratemporal surface of maxilla
12. Lacrimal groove of maxilla
13. Maxillary hiatus and sinus of maxilla
14. Nasal crest of maxilla
15. Orbital surface of maxilla
16. Palatine process of maxilla
17. Zygomatic process of maxilla

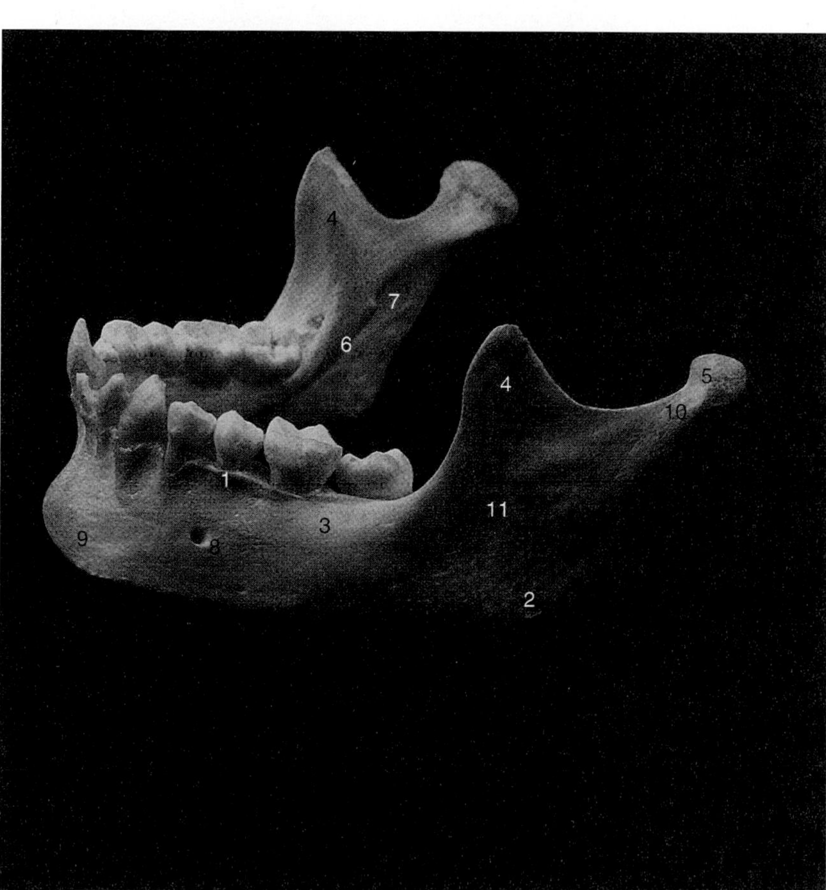

Figure 2-14
Mandible: Left Lateral View

STRUCTURES

1. Alveolar process of mandible
2. Angle of mandible
3. Body of mandible
4. Coronoid process of mandible
5. Head of mandible
6. Lingula of mandible
7. Mandibular foramen of mandible
8. Mental foramen of mandible
9. Mental protuberance of mandible
10. Neck of mandible
11. Ramus of mandible

Note: Together 5 & 10 form the condylar process of mandible

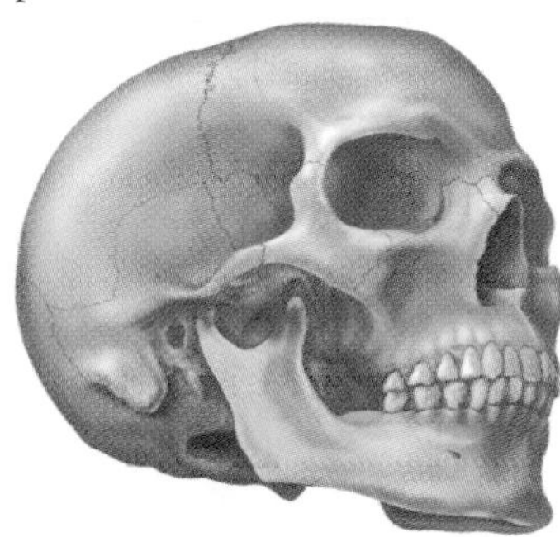

Figure 2-15
Ethmoid Bone:
(A) View from Above, Right, and Behind,
(B) View from Below, Right, and Behind

STRUCTURES

1. Cribriform plate of ethmoid
2. Crista galli of ethmoid
3. Ethmoidal labyrinth (with air cells) of ethmoid
4. Middle nasal concha of ethmoid
5. Orbital plate of ethmoid
6. Perpendicular plate of ethmoid

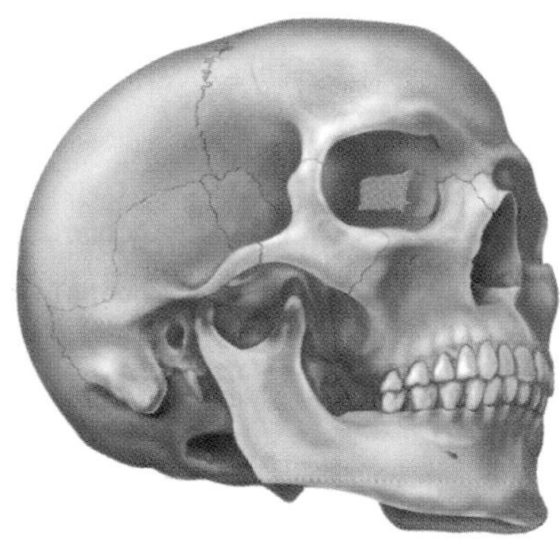

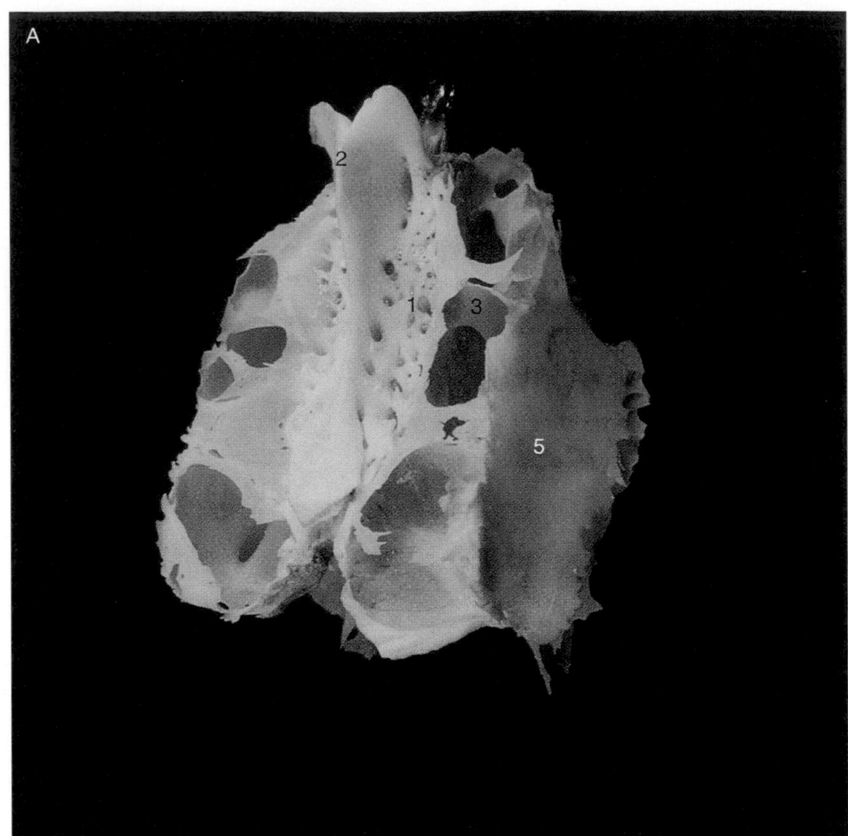

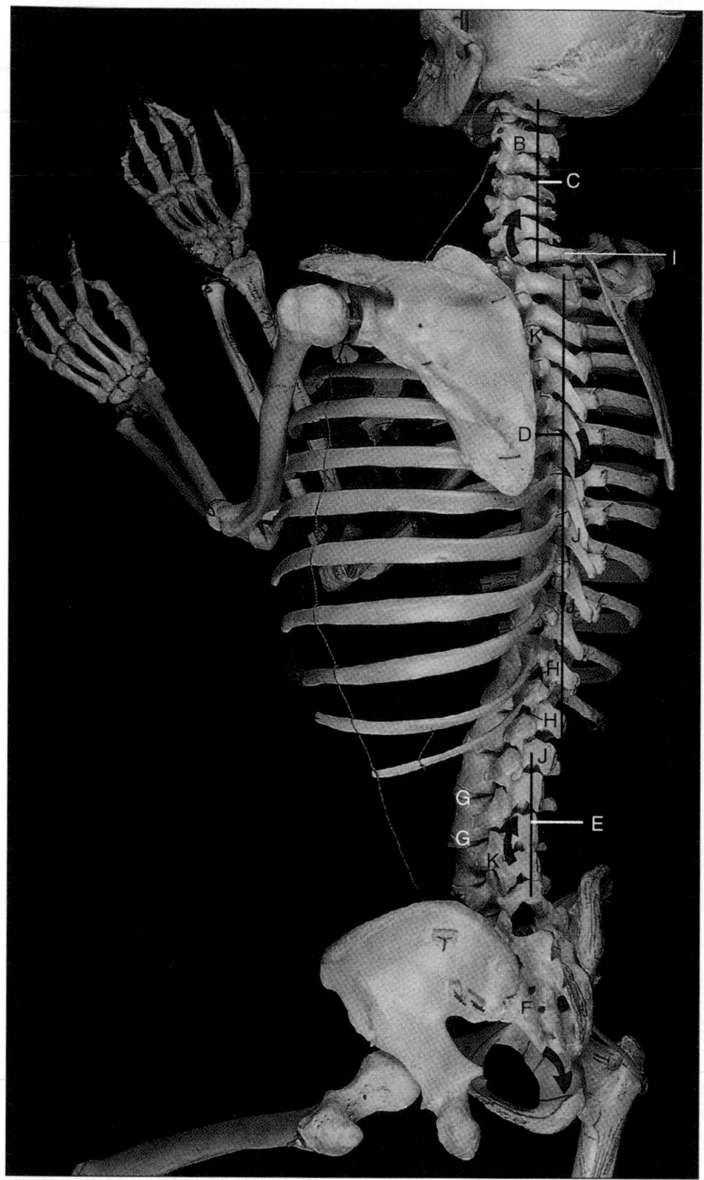

Figure 2-16a
Vertebral Column: View from Left and Behind

STRUCTURES

A. Atlas
B. Axis
C. 7 Cervical vertebrae (Arrow near C—Cervical curvature)
D. 12 Thoracic vertebrae (Arrow near D—Thoracic curvature)
E. 5 Lumbar vertebrae (Arrow near E—Lumbar curvature)
E. 5 Fused sacral vertebrae (Arrow near F—Sacral curvature)
G. Intervertebral disk
H. Intervertebral foramen
I. Mentum nuchae (spinous process of 7th cervical vertebra)
J. Spinous processes
K. Transverse processes

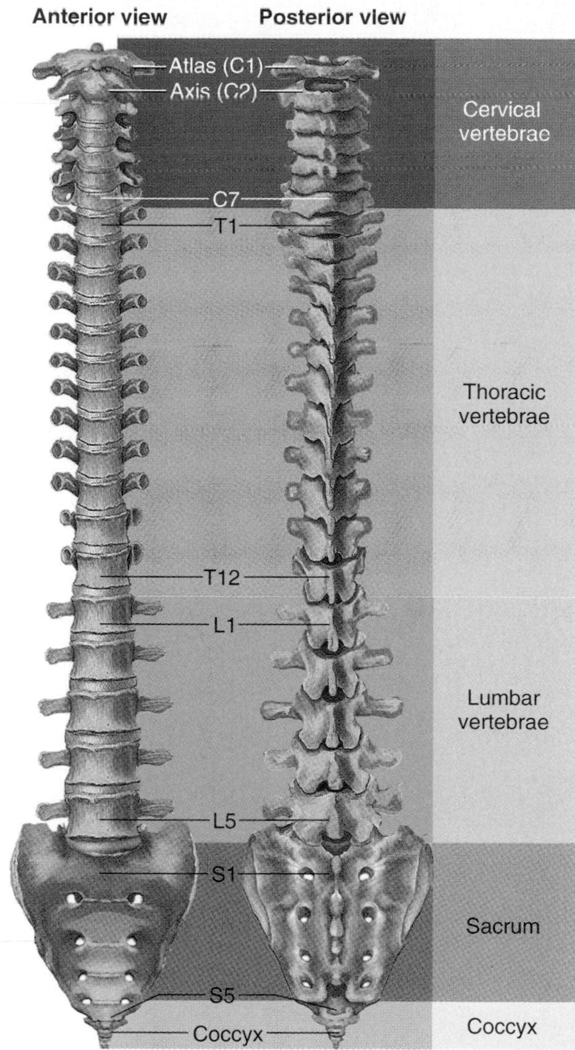

Figure 2-16b
The Vertebral Column, Anterior and Posterior Views

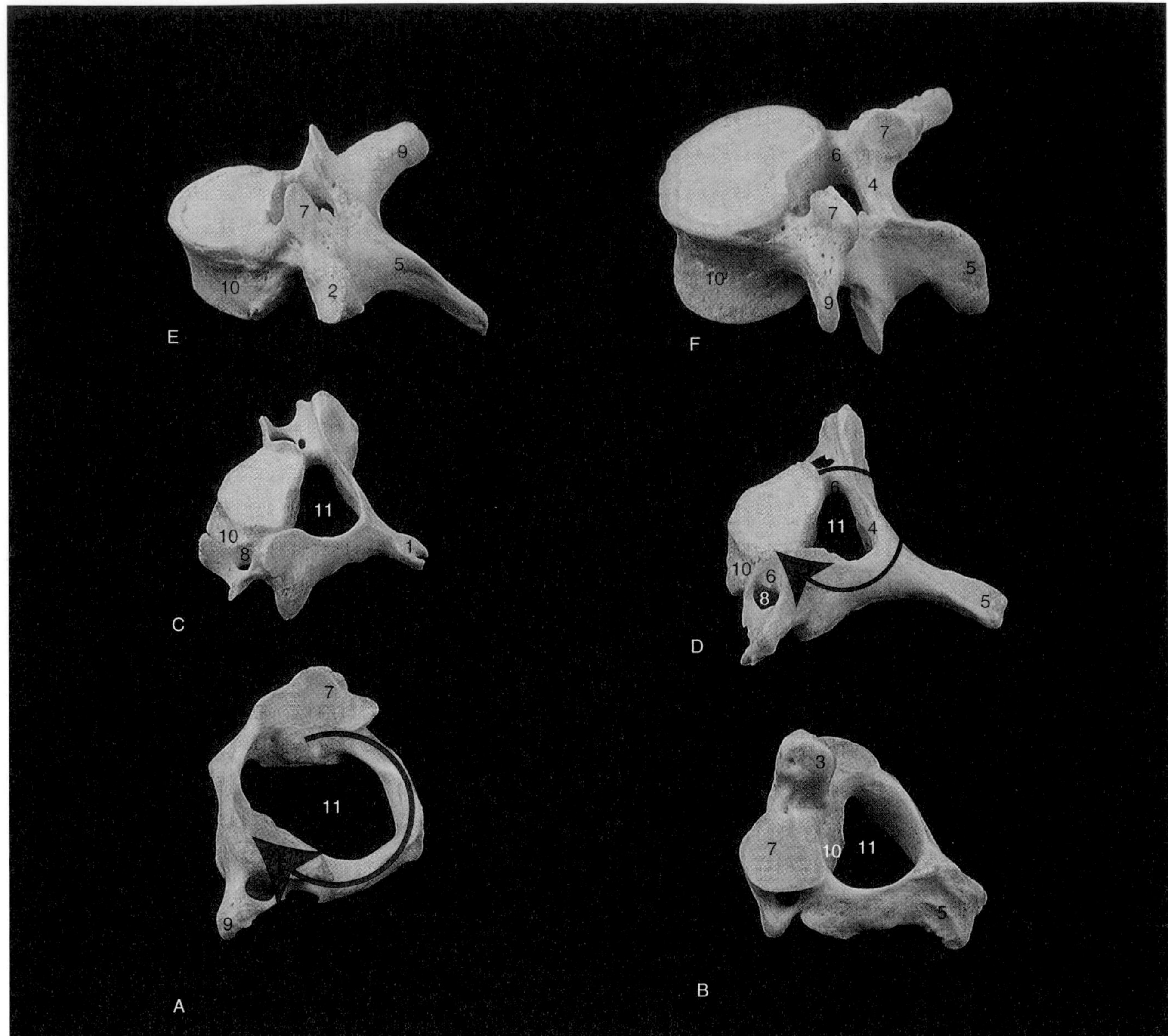

Figure 2-17
Vertebrae

STRUCTURES

A. Atlas
B. Axis
C. Cervical vertebra
D. 7th cervical vertebra
E. Thoracic vertebra
F. Lumbar vertebra

1. Bifid spinous process
2. Costal facet
3. Dens
4. Lamina
5. Monofid spinous process
6. Pedicle
7. Superior articular process and facet
8. Transverse foramen
9. Transverse process
10. Vertebral body
(Heavy arrows—Vertebral arches—comprised of lamina and pedicle)
11. Vertebral foramen

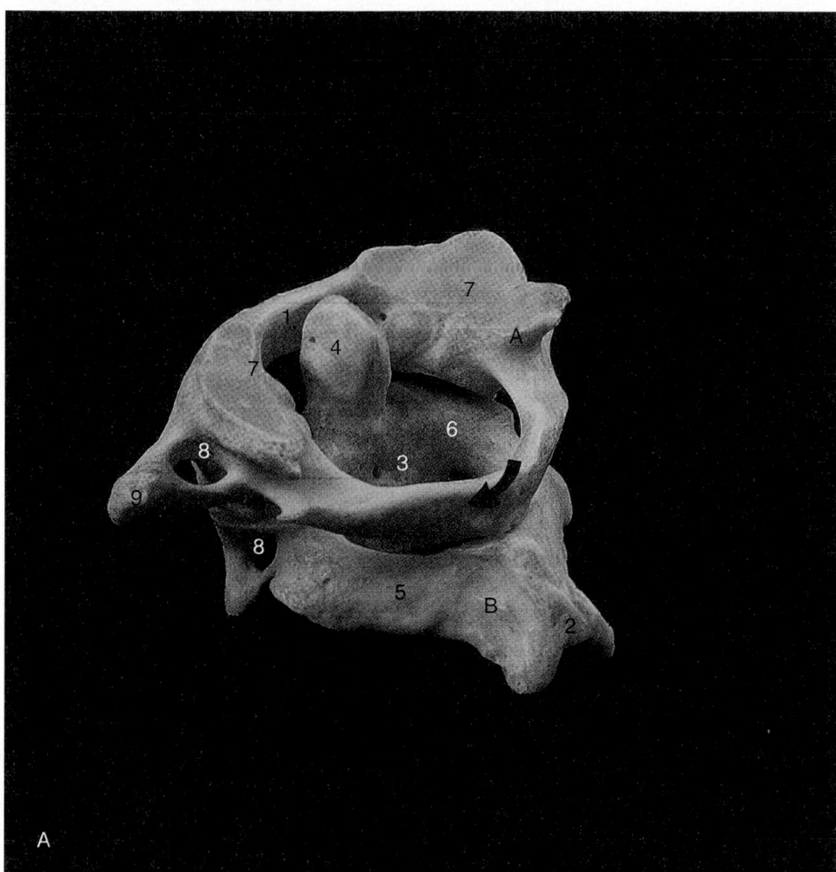

Figure 2-18
(A) Atlas and (B) Axis Articulated

STRUCTURES

1. Anterior arch of atlas
   (Heavy arrow—Posterior arch of atlas)
2. Bifid spinous process
3. Body of axis
4. Dens of axis
5. Lamina of axis
6. Pedicle of axis
7. Superior articular facet
8. Transverse foramen
9. Transverse process

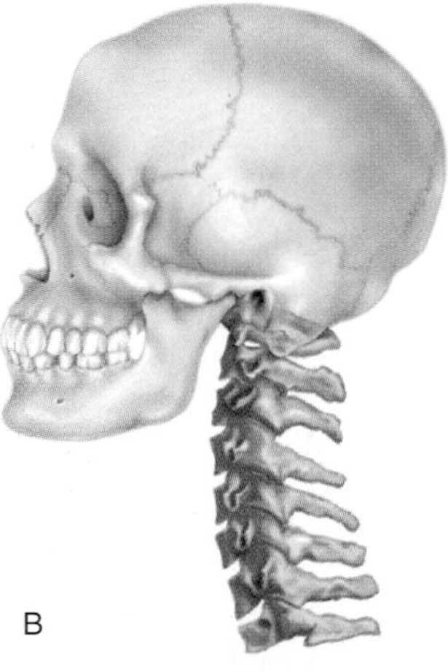

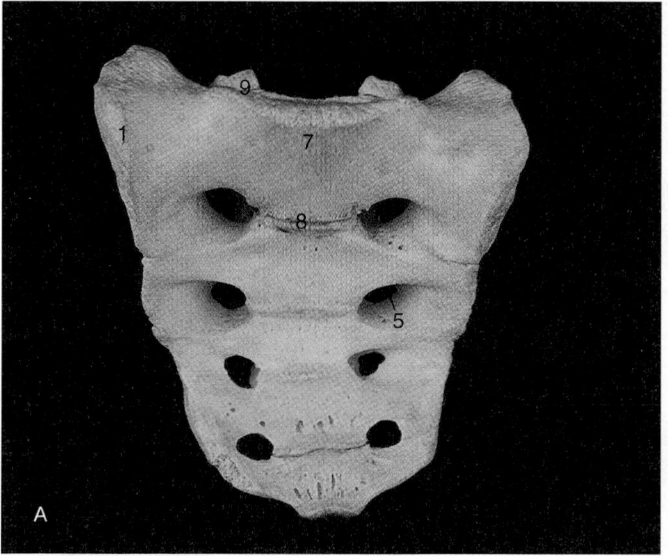

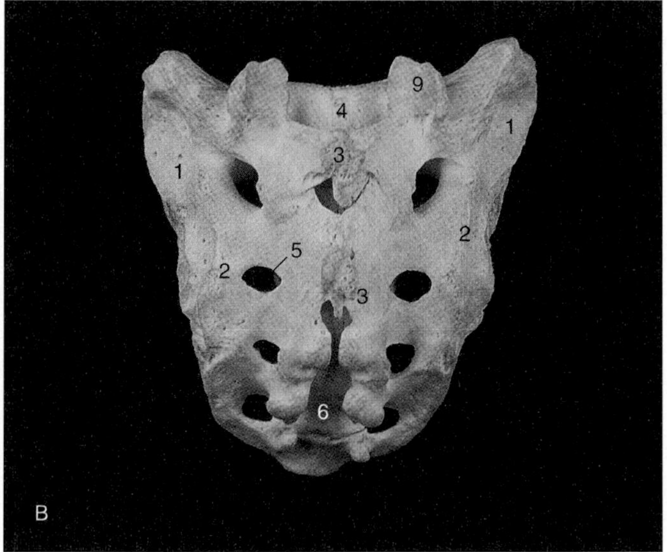

Figure 2-19
Sacrum: (A) Anterior View, (B) Posterior View

STRUCTURES

1. Auricular surface
2. Lateral sacral crest
3. Median sacral crest
4. Sacral canal
5. Sacral foramen
6. Sacral hiatus
7. Sacral promontory
8. Site of fusion of 1st and 2nd sacral vertebrae
9. Superior articular process and facet

Figure 2-20
Sternum and Ribs

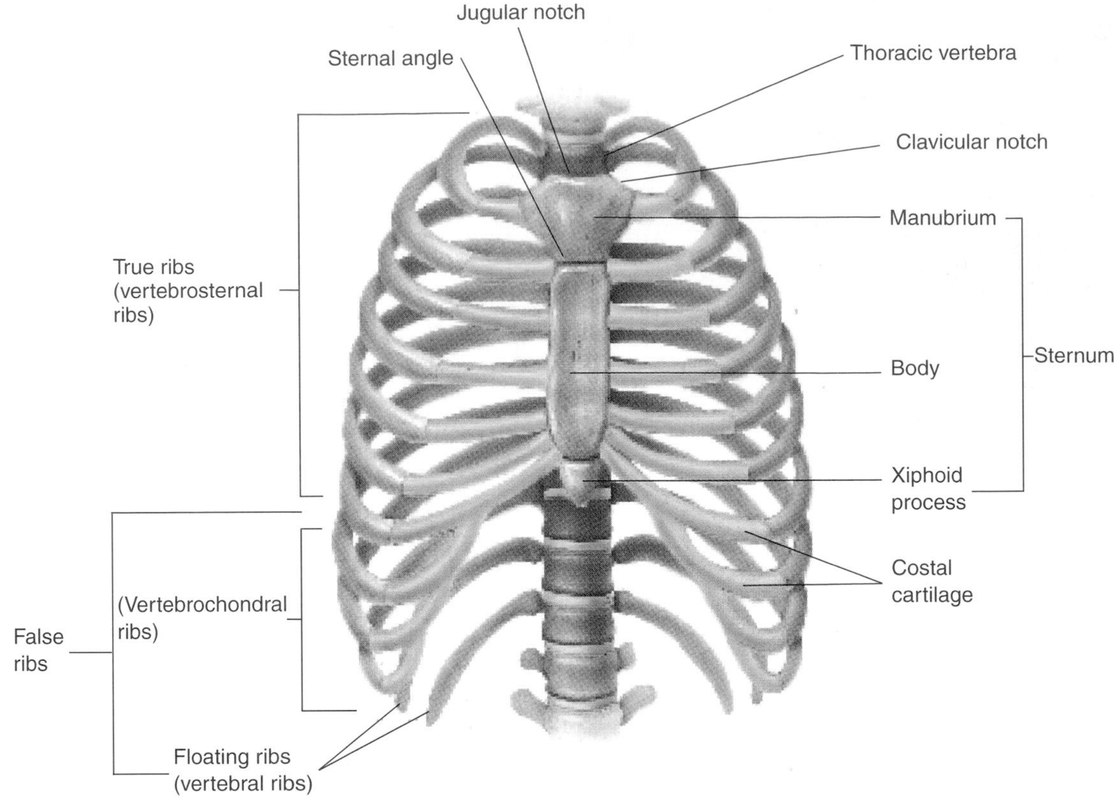

Figure 2-21
Ribs: (A) 1st, (B) 2nd, Right Side, Superior View, (C) 7th, Right Side, Inferior View

STRUCTURES

1. Angle
2. Articular facet
3. Body
4. Costal groove
5. Head
6. Neck
7. Scalene tubercle
8. Serratus anterior tuberosity
9. Site for attachment of levator costa
10. Site for attachment of scalenus medius
11. Subclavian groove
12. Tubercle

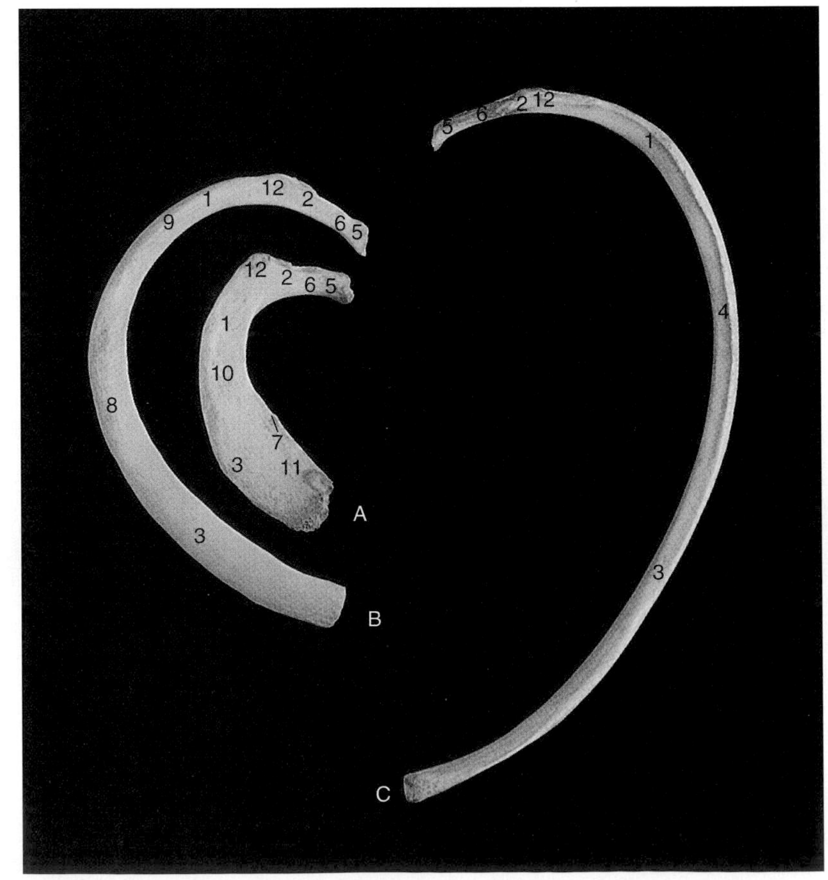

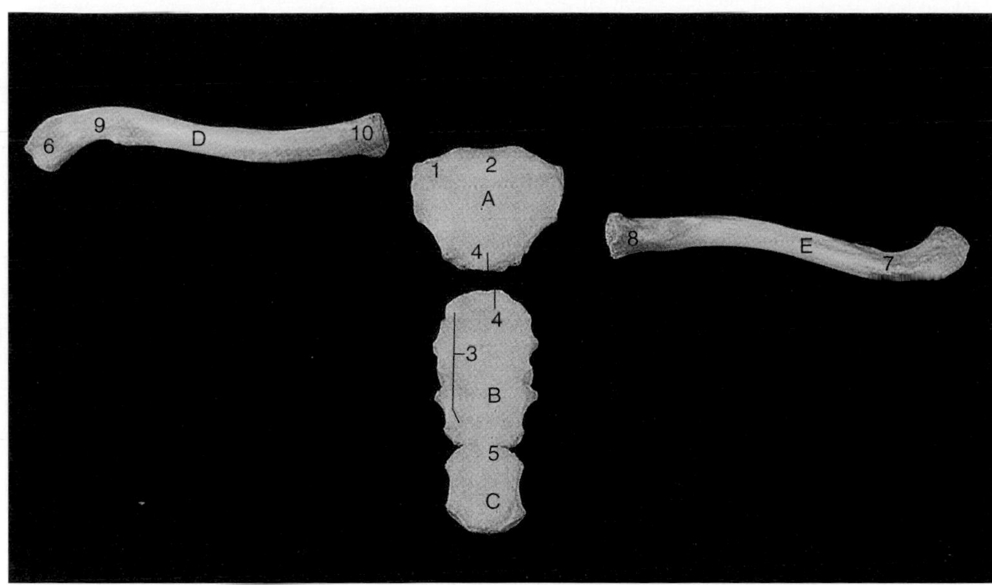

Figure 2-22
Sternum and Clavicles: Sternum from the Front, (D) Right Clavicle, Superior View, (E) Left Clavicle, Inferior View

STRUCTURES

A. Manubrium of sternum
B. Body of sternum
C. Xiphoid process of sternum
1. Clavicular notch
2. Jugular notch
3. Notches for costal cartilages (2–7)
4. Sternal angle and manubriosternal joint
5. Xiphisternal joint
D. Right clavicle
E. Left clavicle
6. Acromial end
7. Conoid tubercle
8. Site for costaclavicular ligament
9. Site for deltoid
10. Sternal end

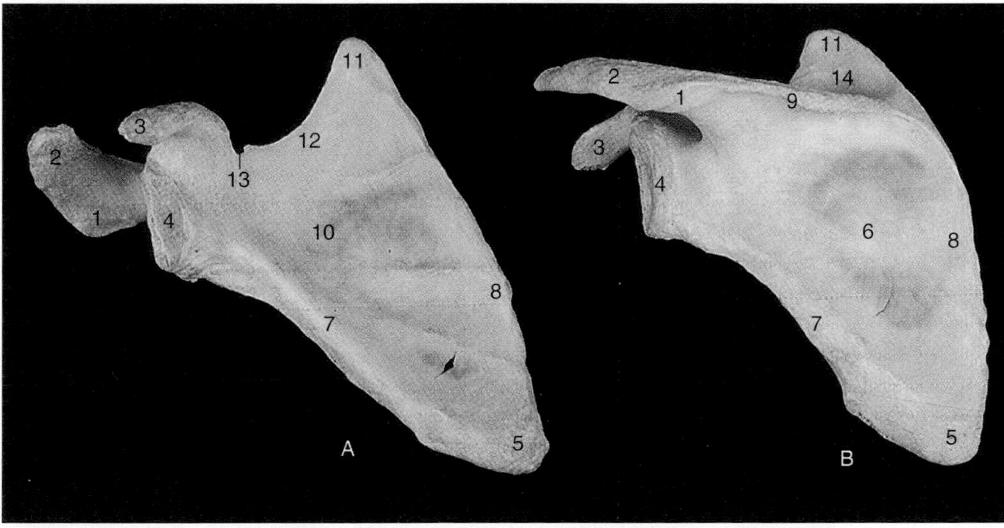

Figure 2-23
Scapulare: (A) Right Scapula, Anterior View, (B) Left Scapula, Posterior View

STRUCTURES

A. Right scapula
B. Left scapula
1. Acromial angle
2. Acromion
3. Coracoid process
4. Glenoid cavity
5. Inferior angle
6. Infraspinous fossa
7. Lateral (axillary) border
8. Medial (vertebral) border
9. Spine of scapula
10. Subscapular fossa
11. Superior angle
12. Superior border
13. Suprascapular notch
14. Supraspinous fossa

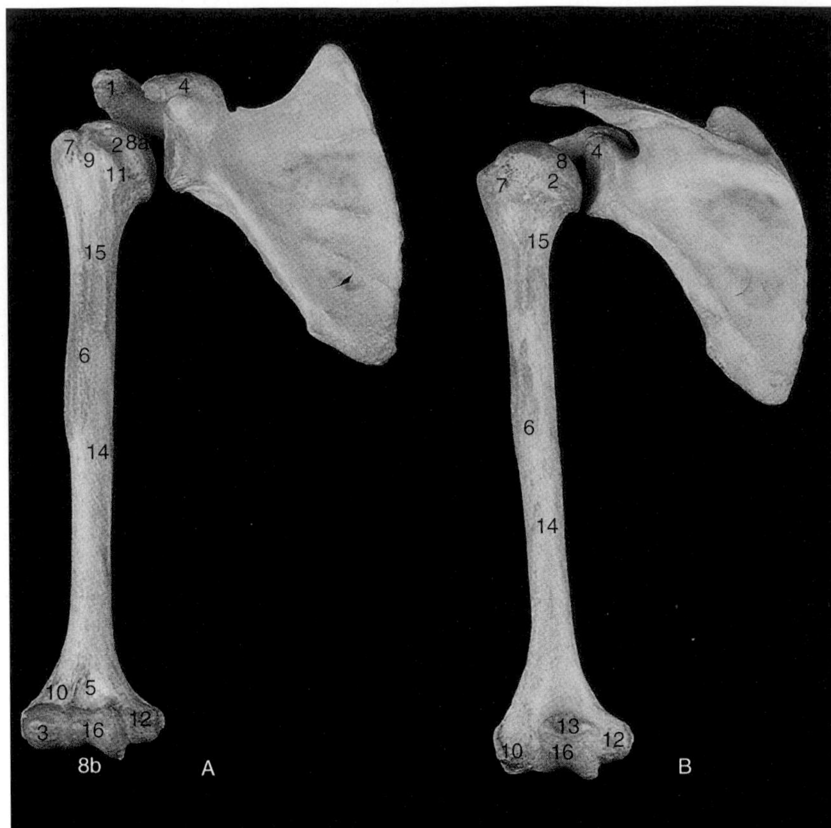

Figure 2-24

Humerus with Scapulae:
(A) Right Humerus, Anterior View,
(B) Left Humerus, Posterior View

STRUCTURES

A. Right humerus and scapula
B. Left humerus and scapula
1. Acromion of scapula
2. Anatomical neck of humerous
3. Capitulum of humerous
4. Coracoid process of scapula
5. Coronoid fossa of humerous
6. Deltoid tuberosity of humerous
7. Greater tubercle of humerous
8. Head (epiphysis) of humerous
   a. proximal
   b. distal
9. Intertubercular sulcus of humerous
10. Lateral epicondyle of humerous
11. Lesser tubercle of humerous
12. Medial epicondyle of humerous
13. Olecranon fossa of humerous
14. Shaft (diaphysis) of humerous
15. Surgical neck of humerous
16. Trochlea of humerous

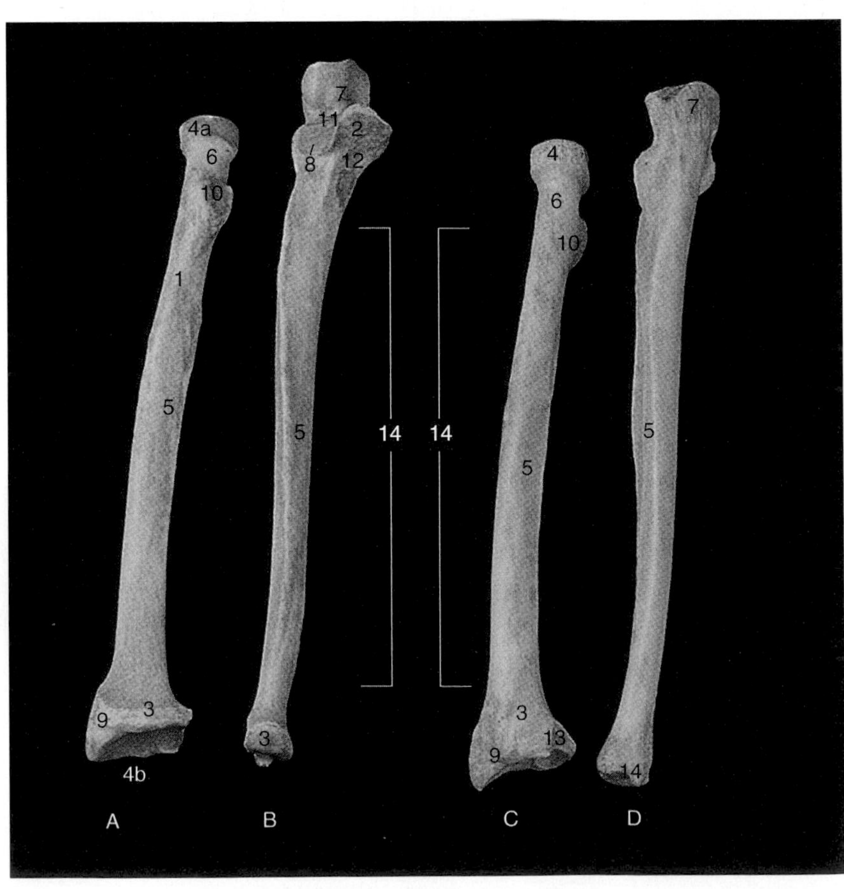

Figure 2-25

Ulna and Radius: Right, Anterior View
Left, Posterior View

STRUCTURES

A. Right radius
B. Right ulna
C. Left radius
D. Left ulna
1. Anterior oblique line of radius
2. Coronoid process of ulna
3. Distal extremity of radius
4. Head of radius
   a. Head of radius
   b. Head of radius
5. Interosseous border
6. Neck of radius
7. Olecranon of ulna
8. Radial notch of ulna
9. Radial styloid process
10. Radial tuberosity of radius
11. Trochlear notch of ulna
12. Tuberosity of ulna
13. Ulnar notch of radius
14. Ulnar styloid process

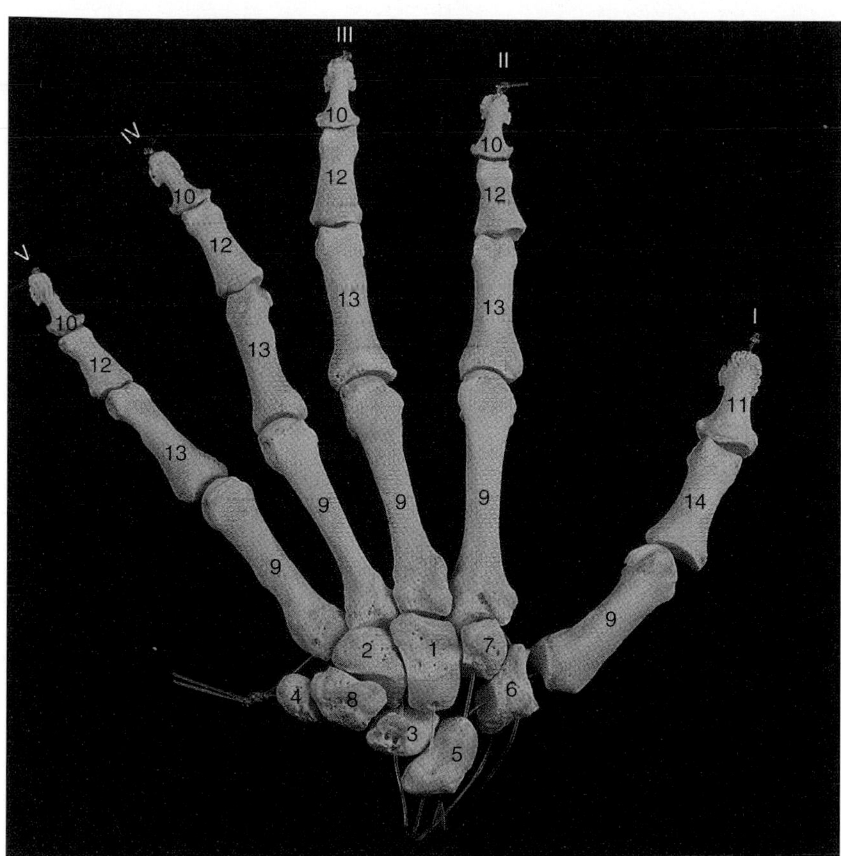

Figure 2-26
Left Hand, Dorsal View

BONES & PROCESSES

CARPALS

1. Capitate
2. Hamate
3. Lunate
4. Pisiform
5. Scaphoid (navicular)
6. Trapezium (greater multangular)
7. Trapezoid (lesser multangular)
8. Triquetrum (triangular)
9. METACARPALS I–V

PHALANGES

10. Distal phalanx of digits II–V
11. Distal phalanx of thumb
12. Middle phalanx of digits II–V
13. Proximal phalanx of digits II–V
14. Proximal phalanx of thumb

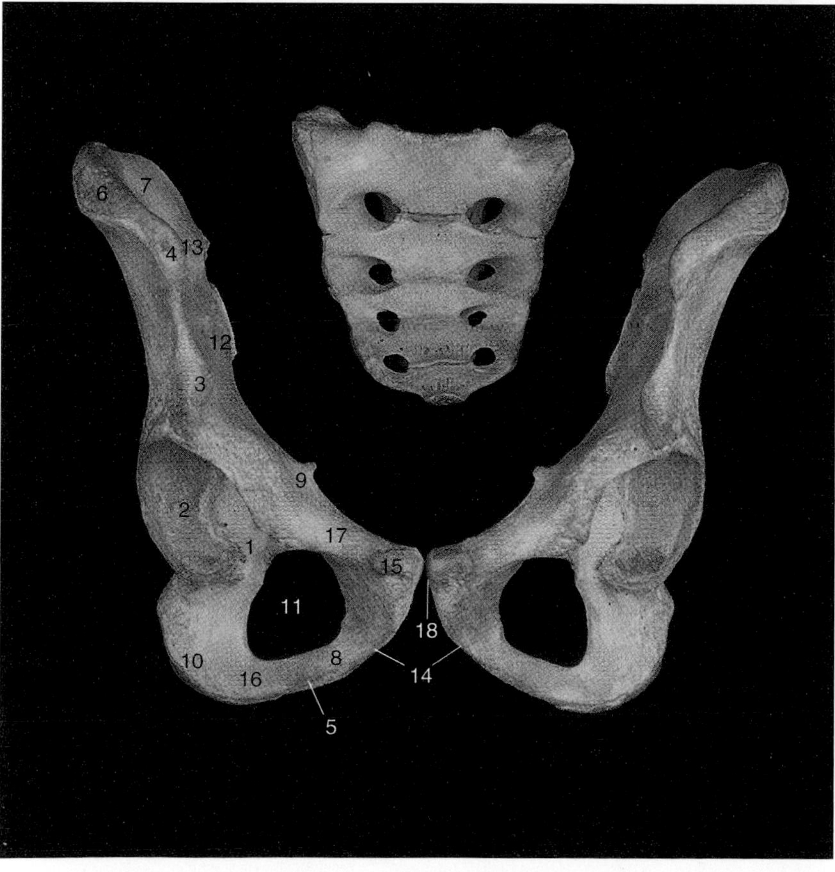

Figure 2-27
Pelvic Bones with Sacrum, Anterior View

STRUCTURES

1. Acetabular notch
2. Acetabulum
3. Anterior inferior iliac spine
4. Anterior superior iliac spine
5. Fusion of ischium and pubis
6. Iliac crest
7. Iliac fossa
8. Inferior pubic ramus
9. Ischial spine
10. Ischial tuberosity
11. Obturator foramen
12. Posterior inferior iliac spine
13. Posterior superior iliac spine
14. Pubic arch
15. Pubic tubercle
16. Ramus of ischium
17. Superior pubic ramus
18. Symphysis pubis (location indicated)

Figure 2-28
Pelvic Bones:
(A) Right, Lateral View,
(B) Left, Medial View

STRUCTURES

1. Acetabular notch
2. Acetabulum
3. Anterior inferior iliac spine
4. Anterior superior iliac spine
5. Arcuate line of ilium
6. Auricular surface (for sacroiliac joint)
7. Fusion of ischium and pubis
8. Greater sciatic notch
9. Iliac crest
10. Iliac fossa
11. Iliopubic eminence
12. Inferior pubic ramus
13. Ischial spine
14. Ischial tuberosity
15. Lesser sciatic notch
16. Obturator foramen
17. Posterior inferior iliac spine
18. Posterior superior iliac spine
19. Ramus of ischium
20. Superior pubic ramus
21. Symphysis pubis (location shown)

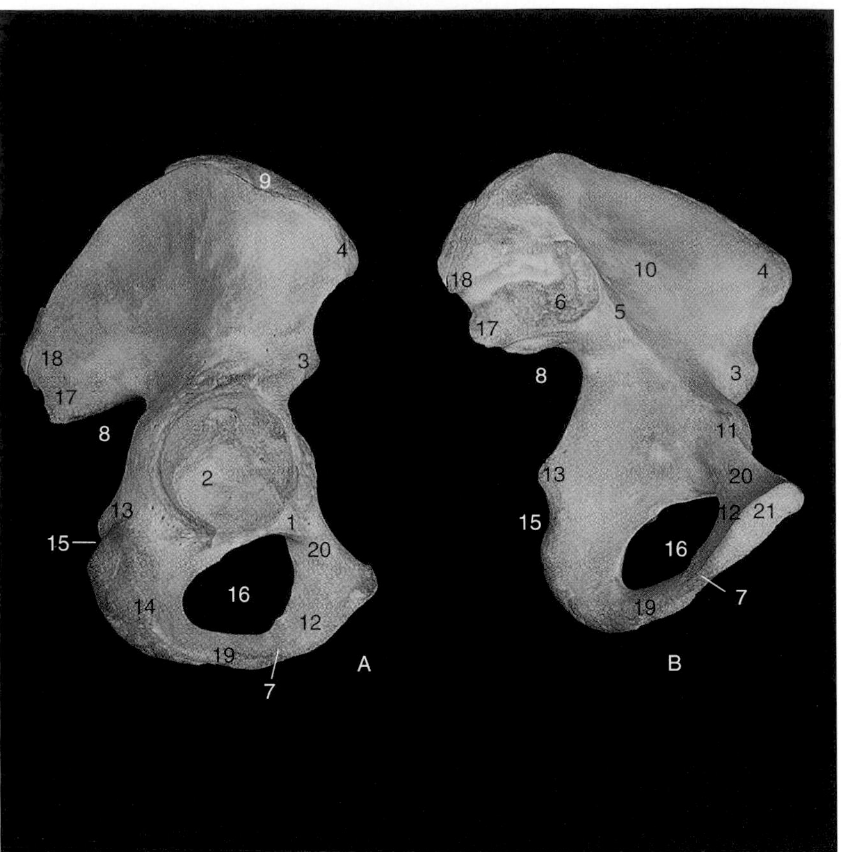

Figure 2-29
Femur: (A) Right Femur, Anterior View,
(B) Left Femur, Posterior View

STRUCTURES

A. Right femur
B. Left femur
1. Neck
2. Gluteal tuberosity
3. Greater trochanter
4. Head (both proximal and distal)
5. Intercondylar fossa
6. Intertrochanteric crest
7. Intertrochanteric line
8. Lateral condyle
9. Lateral epicondyle
10. Lesser trochanter
11. Linea aspera
12. Medial condyle
13. Medial epicondyle
14. Patellar surface
15. Popliteal surface
16. Shaft (diaphysis)
17. Site for attachment of anterior cruciate ligament
18. Site for attachment of posterior cruciate ligament
19. Spiral line

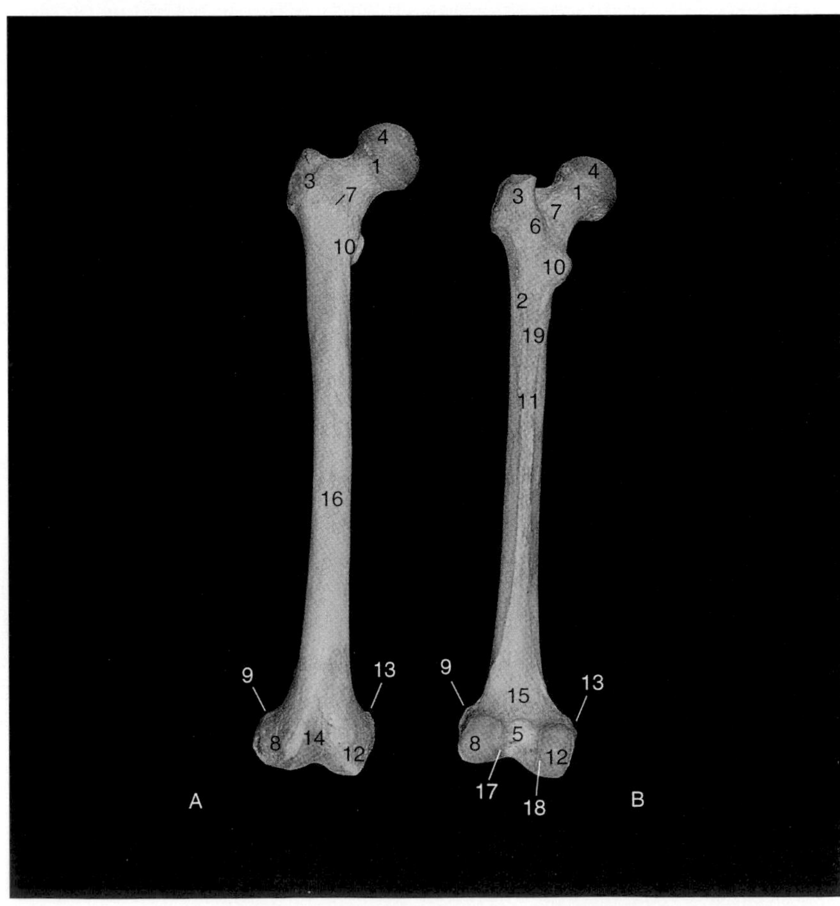

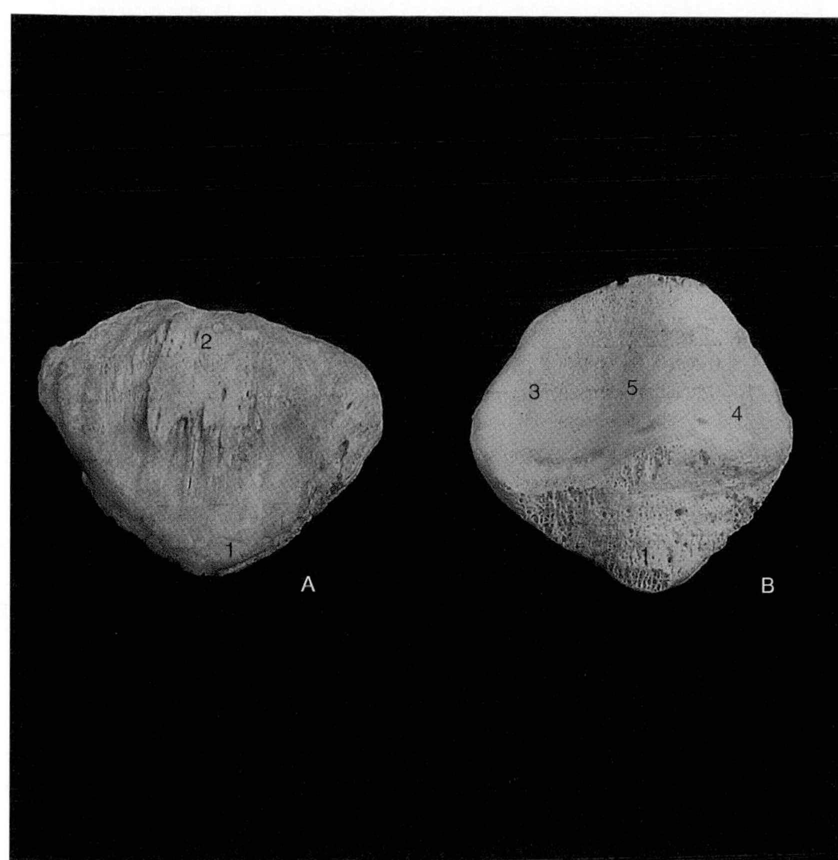

Figure 2-30
Patellae: (A) Right Patella, Anterior Surface, (B) Left Patella, Posterior (Articular) Surface

STRUCTURES

A. Right patella
B. Left patella
1. Apex (site for attachment of patellar ligament)
2. Base
3. Facet for lateral condyle of femur
4. Facet for medial condyle of femur
5. Vertical ridge

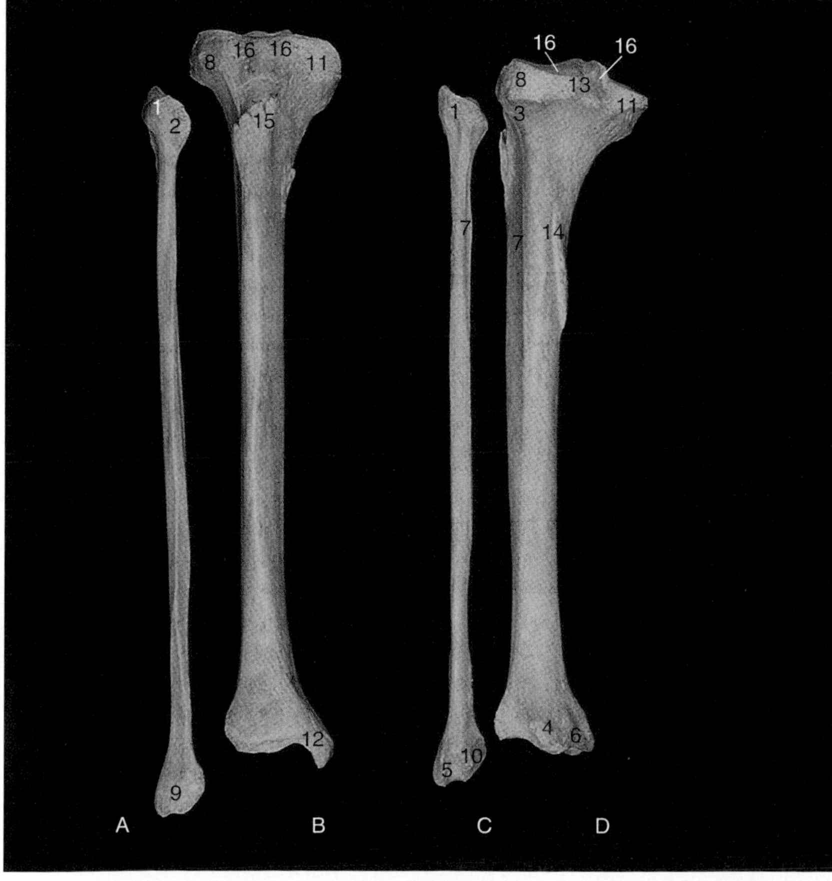

Figure 2-31
Tibia and Fibula: (A,B) Right Tibia and Fibula, Anterior View, (C,D) Left Tibia and Fibula, Posterior View

STRUCTURES

A. Right fibula
B. Right tibia
C. Left fibula
D. Left tibia
1. Apex of fibula
2. Articular facet for tibia
3. Fibular articular facet of tibia
4. Groove for flexor hallicus longus tendon
5. Groove for peroneus brevis tendon
6. Groove for tibialis posterior tendon
7. Interosseous border
8. Lateral condyle of tibia
9. Lateral malleolus of fibula
10. Malleolar fossa of fibula
11. Medial condyle of tibia
12. Medial malleolus of tibia
13. Site for attachment of posterior cruciate ligament
14. Soleal line of tibia
15. Tibial tuberosity
16. Tubercles of intercondylar eminence of tibia

Figure 2-32
Right Foot: Dorsal (Superior) View

BONES & PROCESSES

TARSALS

1. Calcaneus
2. Cuboid
3. Groove for flexor hallucis longus tendon
4. Head of talus
5. Intermediate cuneiform
6. Lateral cuneiform
7. Lateral tubercle of talus
8. Medial cuneiform
9. Medial tubercle of talus
10. Navicular
11. Navicular tuberosity
12. Neck of talus
13. Trochlea (articular surface) of talus
14. Metatarsals I–V

PHALANGES

15. Distal phalanx of digits II–V
16. Distal phalanx of great toe
17. Middle phalanx of digits II–V
18. Proximal phalanx of digits II–V
19. Proximal phalanx of great toe
20. Site for attachment of Achilles tendon

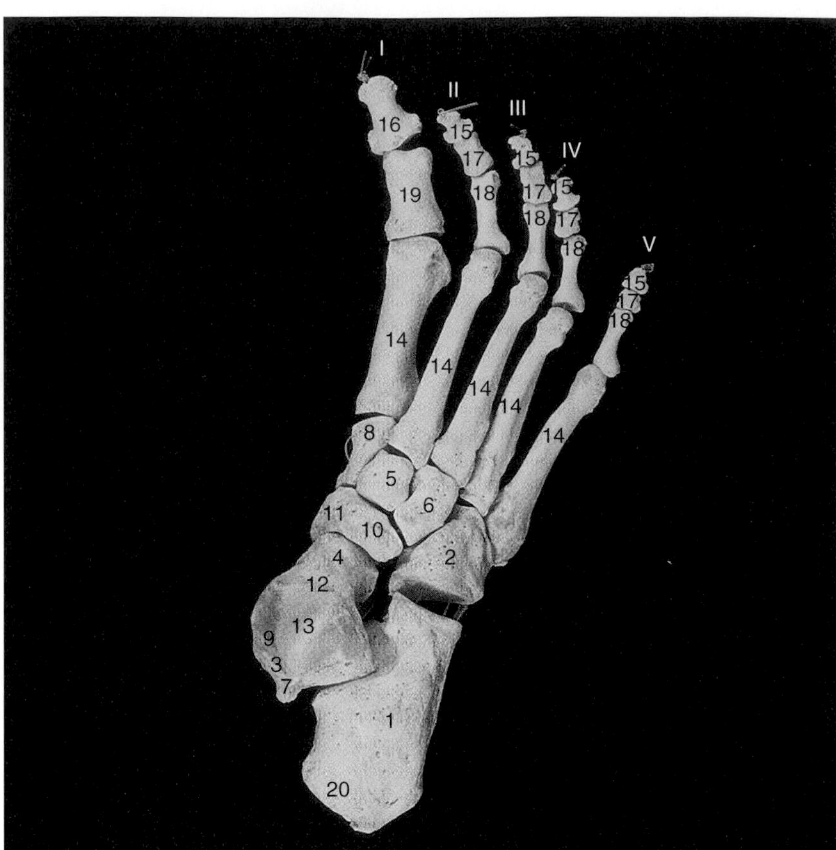

Figure 2-33
Typical Long Bone Structure

STRUCTURES

1. Cancellous (spongy) bone
2. Compact bone
3. Head (epiphysis)
   a. proximal
   b. distal
4. Medullary (marrow) cavity
5. Neck (growth plate, metaphysis, epiphyseal plate)
6. Shaft (diaphysis)

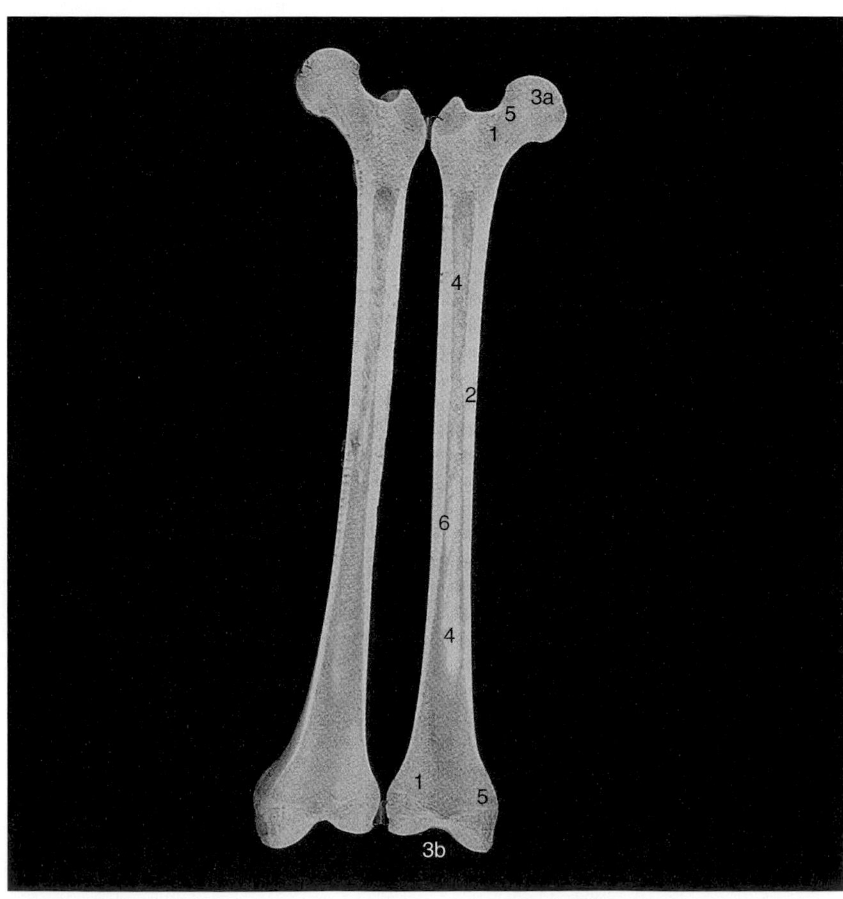

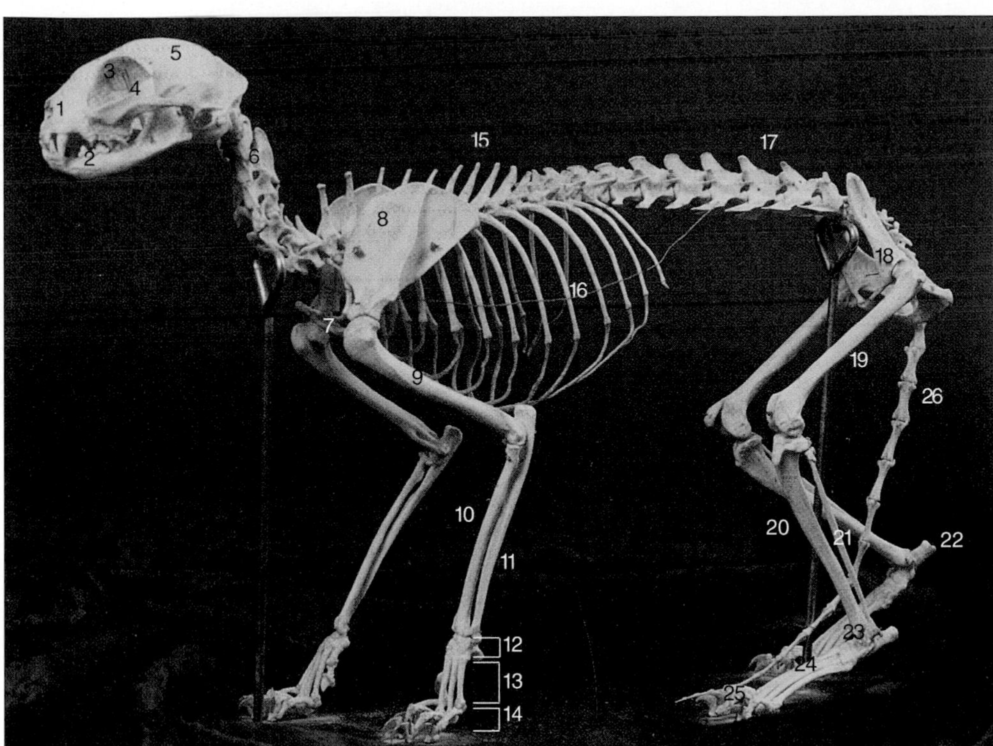

Figure 2-34
The Cat Skeleton

1. Maxilla
2. Mandible
3. Orbit
4. Zygomatic arch
5. Cranium
6. Cervical vertebrae (7)
7. Sternum
8. Scapula
9. Humerus
10. Radius
11. Ulna
12. Carpal bones
13. Metacarpal bones
14. Phalanges
15. Thoracic vertebrae (13)
16. Ribs
17. Lumbar vertebrae (7)
18. Pelvis
19. Femur
20. Tibia
21. Fibula
22. Calcaneus
23. Tarsal bones
24. Metatarsal bones
25. Phalanges
26. Caudal vertebrae (21–25)

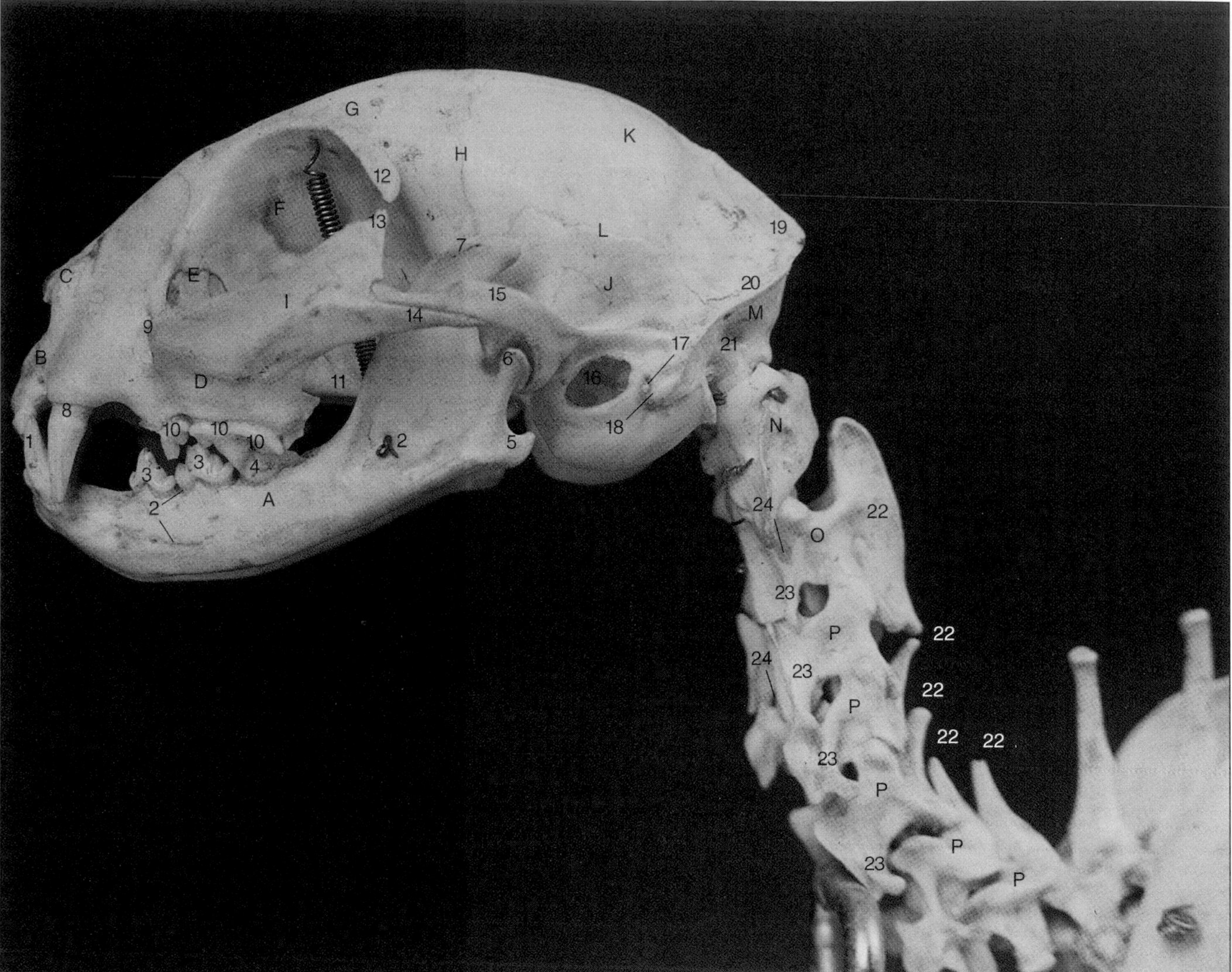

Figure 2-35
Cat Skull, Left Lateral View

A. Mandible
  1. Lower canine tooth
  2. Mental foramina
  3. Lower premolar teeth
  4. Lower molar teeth
  5. Angular process
  6. Condylar process
  7. Coronoid process
B. Incisive bone
C. Nasal bone
D. Maxilla
  8. Upper canine tooth
  9. Infraorbital foramen
  10. Upper premolar tooth
  11. Upper molar tooth
E. Lacrimal bone and fossa
F. Orbit
G. Frontal bone
  12. Zygomatic process of frontal bone
H. Coronal suture
I. Zygomatic bone
  13. Frontal process of zygomatic
  14. Temporal process of zygomatic
J. Temporal bone
  15. Zygomatic process of temporal bone
  16. External auditory meatus
  17. Stylomastoid foramen
  18. Mastoid process
K. Parietal bone
L. Squamosal suture
M. Occipital bone
  19. External occipital protuberance
  20. Nuchal crest
  21. Occipital condyle
N. Atlas
O. Axis
P. Cervical vertebrae (3–7)
  22. Spinous process
  23. Transverse process
  24. Transverse foramen

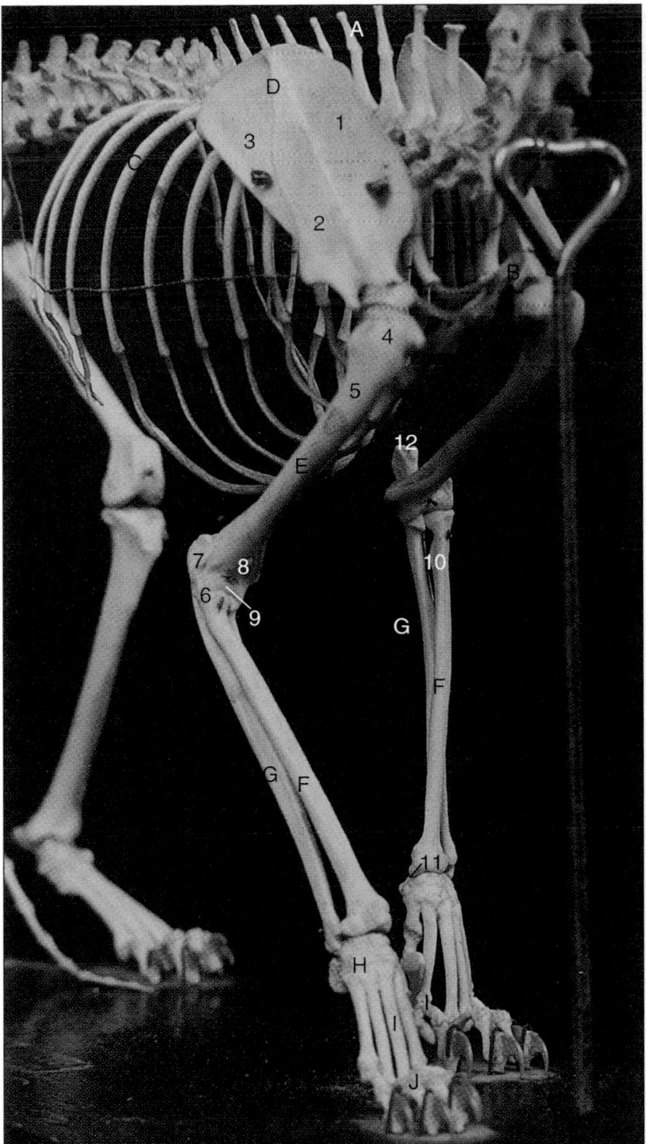

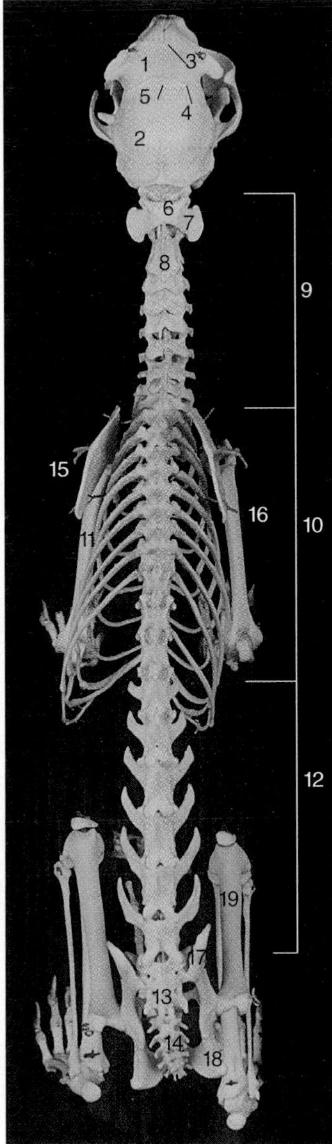

Figure 2-37
Axial Skeleton of Cat, Dorsal View

1. Frontal bone
2. Parietal bone
3. Sagittal suture
4. Coronal suture
5. Bregma
6. Atlas
7. Transverse process (wing) of atlas
8. Axis
9. Cervical vertebrae (7)
10. Thoracic vertebrae (13)
11. Ribs
12. Lumbar vertebrae (7)
13. Sacral vertebrae (3)
14. Caudal vertebrae (21–25)
15. Scapula
16. Humerus
17. Ilium
18. Ischium
19. Femur

Figure 2-36
Cat Skeleton, Front Right Lateral View

A. Vertebral spinous process
B. Sternum
C. Ribs
D. Scapula
   1. Supraspinous fossa
   2. Acromial spine
   3. Infraspinous fossa
E. Humerus
   4. Head
   5. Deltoid tuberosity
   6. Trochlea
   7. Lateral epicondyle
   8. Medial epicondyle
   9. Radial fossa
F. Radius
   10. Radial tuberosity
   11. Styloid process
G. Ulna
   12. Olecranon process
H. Carpal bones
I. Metacarpal bones
J. Phalanges

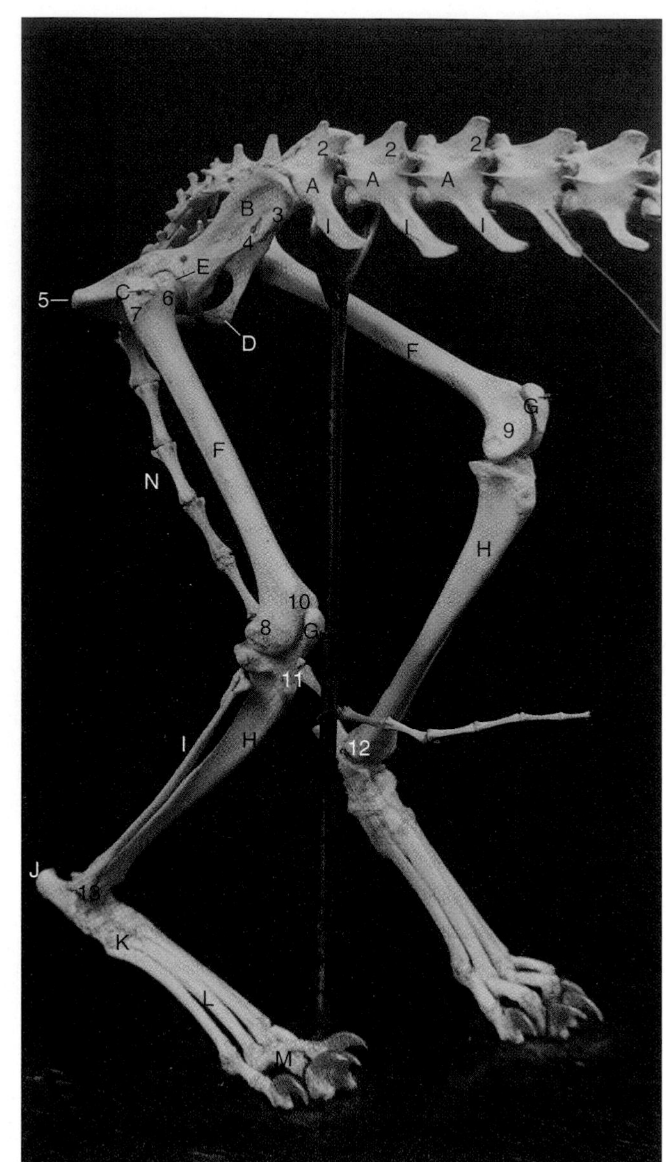

Figure 2-38
Cat Skeleton, Right Lateral View

A. Lumbar vertebrae
  1. Transverse processes
  2. Spinous processes
B. Ilium
  3. Cranial ventral iliac spine
  4. Caudal ventral iliac spine
C. Ischium
  5. Ischial tuberosity
D. Pubis
E. Acetabulum
F. Femur
  6. Head
  7. Greater trochanter
  8. Lateral condyle
  9. Medial condyle
  10. Trochlea
G. Patella
H. Tibia
  11. Tibial tuberosity
  12. Medial malleolus
I. Fibula
  13. Lateral malleolus
J. Calcaneus
K. Tarsal bones
L. Metatarsal bones
M. Phalanges
N. Caudal vertebrae

CHAPTER 3

Human Muscular Anatomy

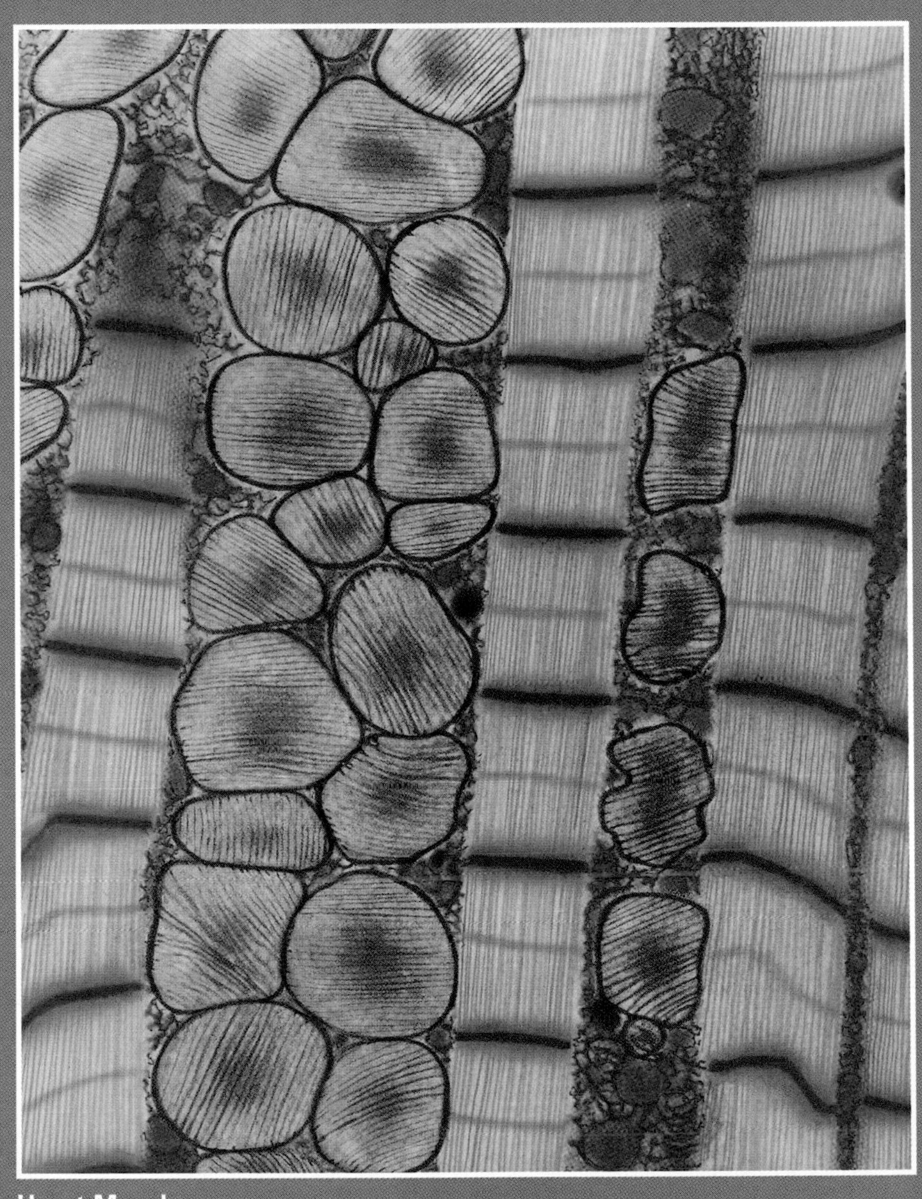

Heart Muscle.
Colored transmission electron micrograph

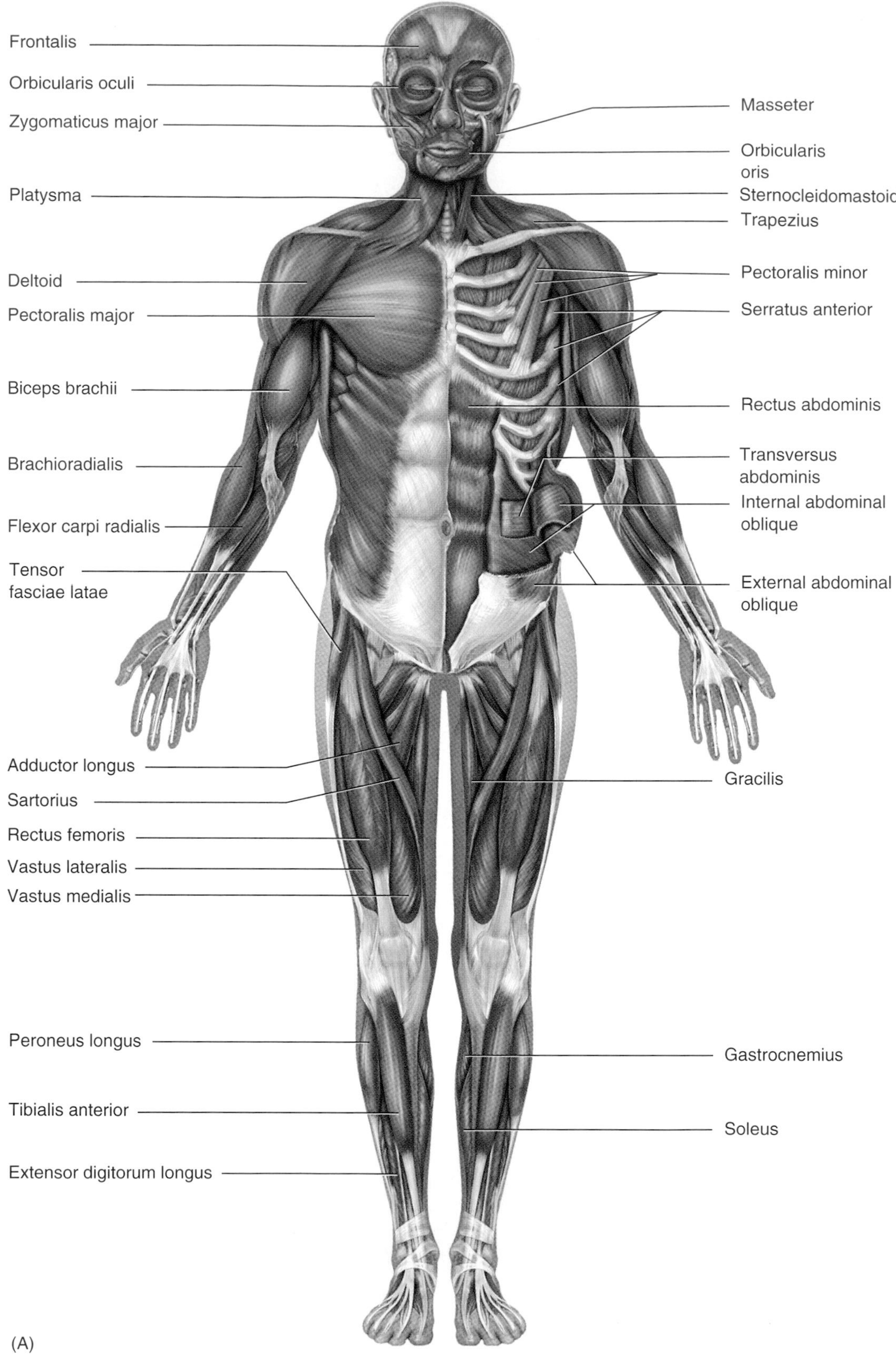
Frontalis
Orbicularis oculi
Zygomaticus major
Platysma
Deltoid
Pectoralis major
Biceps brachii
Brachioradialis
Flexor carpi radialis
Tensor fasciae latae
Adductor longus
Sartorius
Rectus femoris
Vastus lateralis
Vastus medialis
Peroneus longus
Tibialis anterior
Extensor digitorum longus
Masseter
Orbicularis oris
Sternocleidomastoid
Trapezius
Pectoralis minor
Serratus anterior
Rectus abdominis
Transversus abdominis
Internal abdominal oblique
External abdominal oblique
Gracilis
Gastrocnemius
Soleus
(A)

Figure 3-1
Superficial Skeletal Muscles; (A) Anterior view

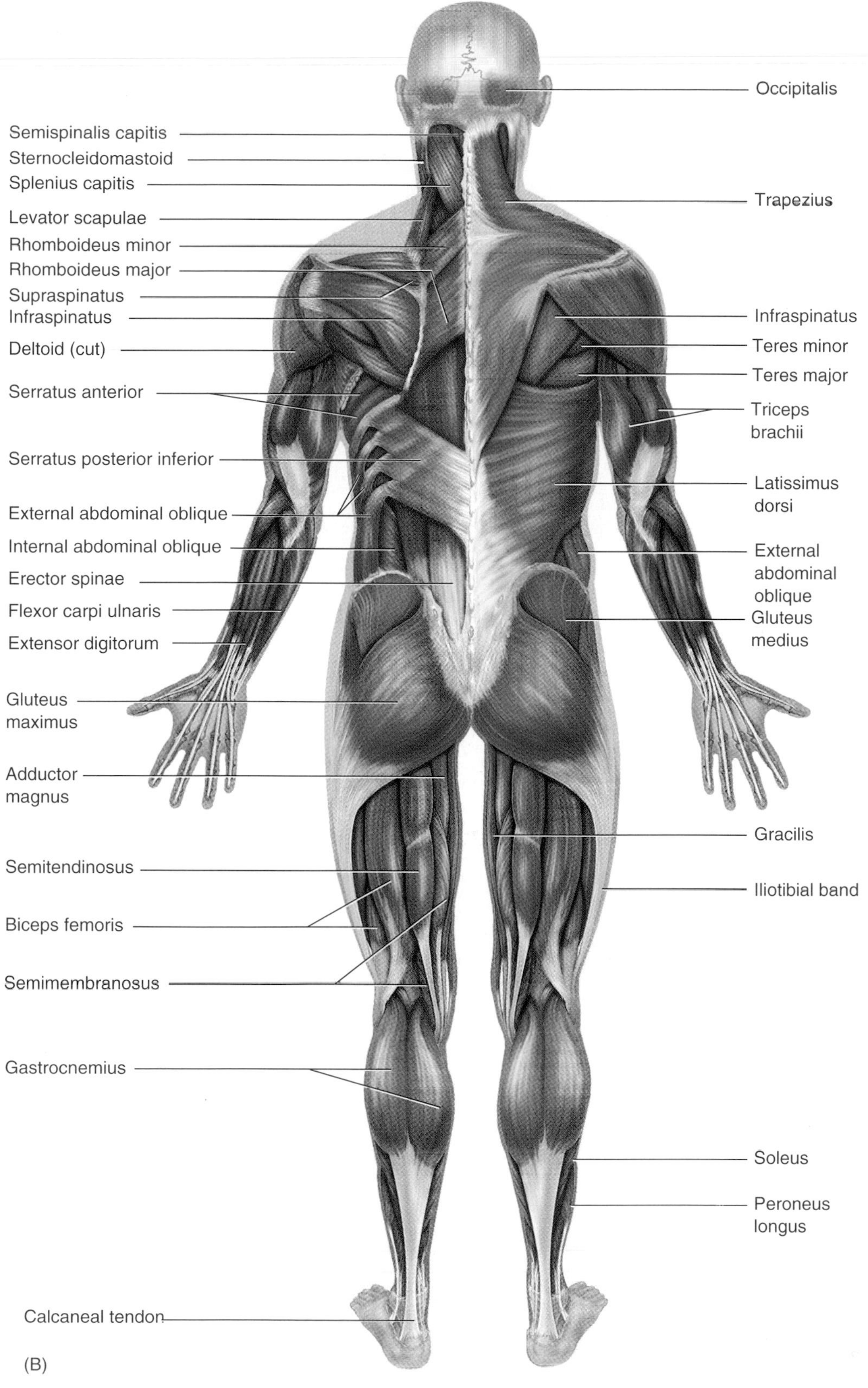
Occipitalis
Semispinalis capitis
Sternocleidomastoid
Splenius capitis
Trapezius
Levator scapulae
Rhomboideus minor
Rhomboideus major
Supraspinatus
Infraspinatus
Infraspinatus
Deltoid (cut)
Teres minor
Teres major
Serratus anterior
Triceps brachii
Serratus posterior inferior
Latissimus dorsi
External abdominal oblique
Internal abdominal oblique
External abdominal oblique
Erector spinae
Flexor carpi ulnaris
Gluteus medius
Extensor digitorum
Gluteus maximus
Adductor magnus
Gracilis
Semitendinosus
Iliotibial band
Biceps femoris
Semimembranosus
Gastrocnemius
Soleus
Peroneus longus
Calcaneal tendon
(B)

Figure 3-1—cont'd.
(B) Posterior view

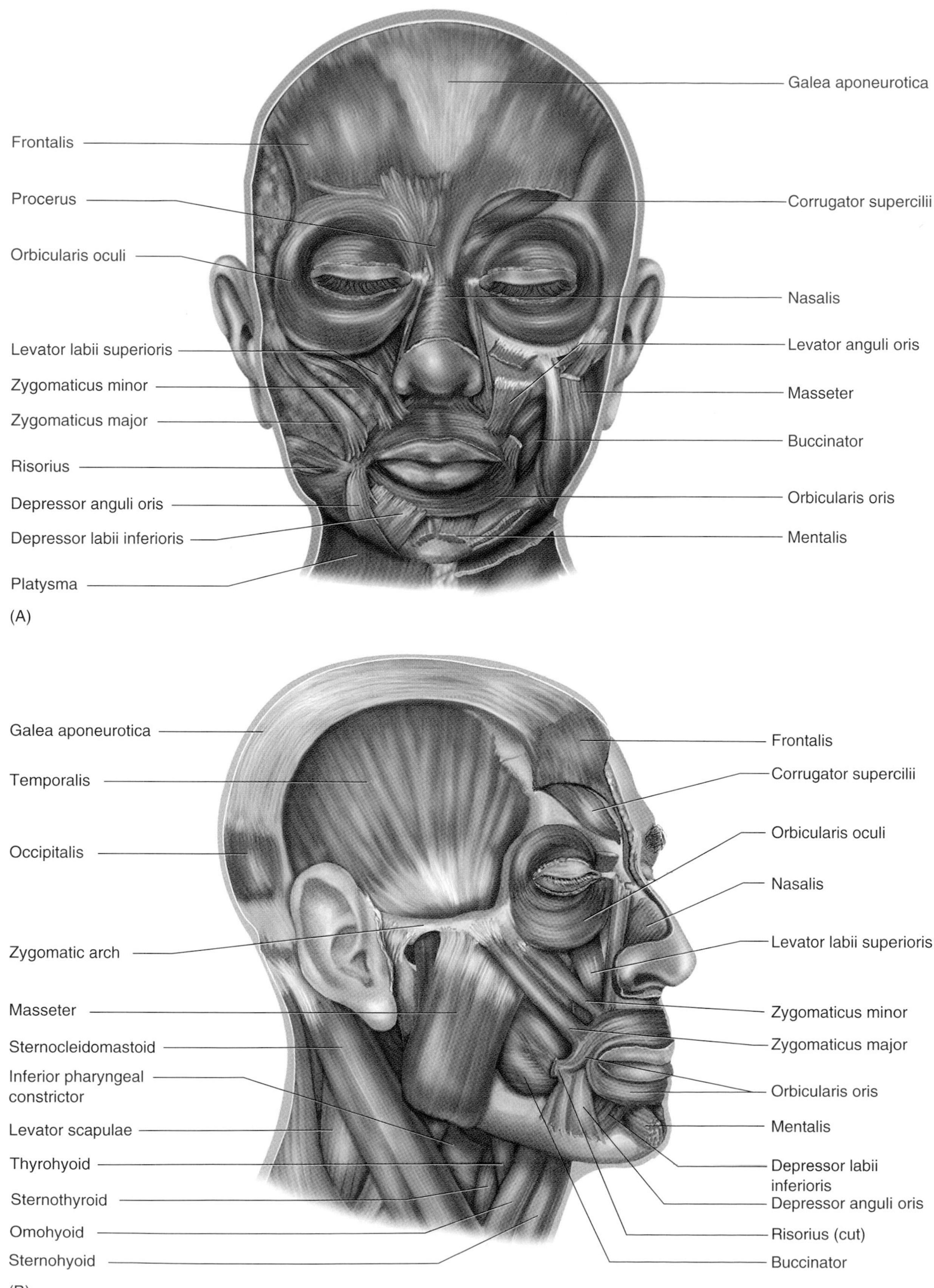
Galea aponeurotica
Frontalis
Procerus
Corrugator supercilii
Orbicularis oculi
Nasalis
Levator labii superioris
Levator anguli oris
Zygomaticus minor
Masseter
Zygomaticus major
Buccinator
Risorius
Orbicularis oris
Depressor anguli oris
Depressor labii inferioris
Mentalis
Platysma
(A)
Galea aponeurotica
Frontalis
Temporalis
Corrugator supercilii
Orbicularis oculi
Occipitalis
Nasalis
Levator labii superioris
Zygomatic arch
Masseter
Zygomaticus minor
Sternocleidomastoid
Zygomaticus major
Inferior pharyngeal constrictor
Orbicularis oris
Levator scapulae
Mentalis
Thyrohyoid
Depressor labii inferioris
Sternothyroid
Depressor anguli oris
Omohyoid
Risorius (cut)
Sternohyoid
Buccinator
(B)

Figure 3-2
(A) Muscles of the head, anterior view, (B) Muscles of the head, lateral view

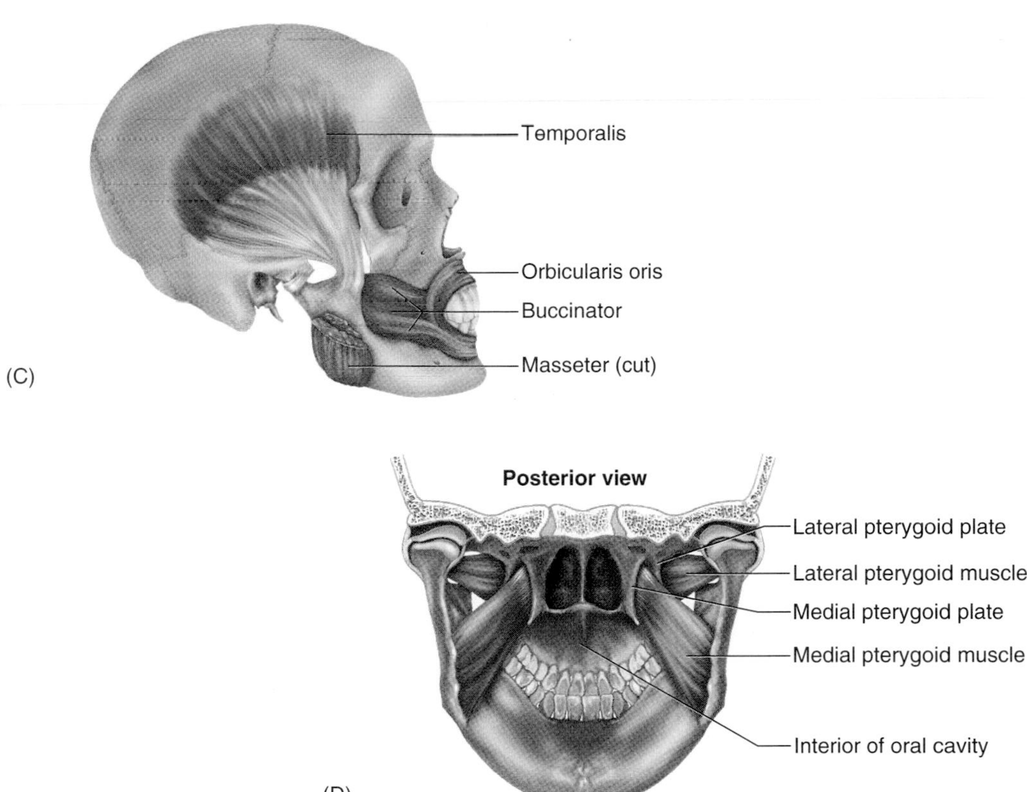

Figure 3-2—cont'd.
(C) Muscles of mastication, (D) The lateral and medial pterygoid muscles

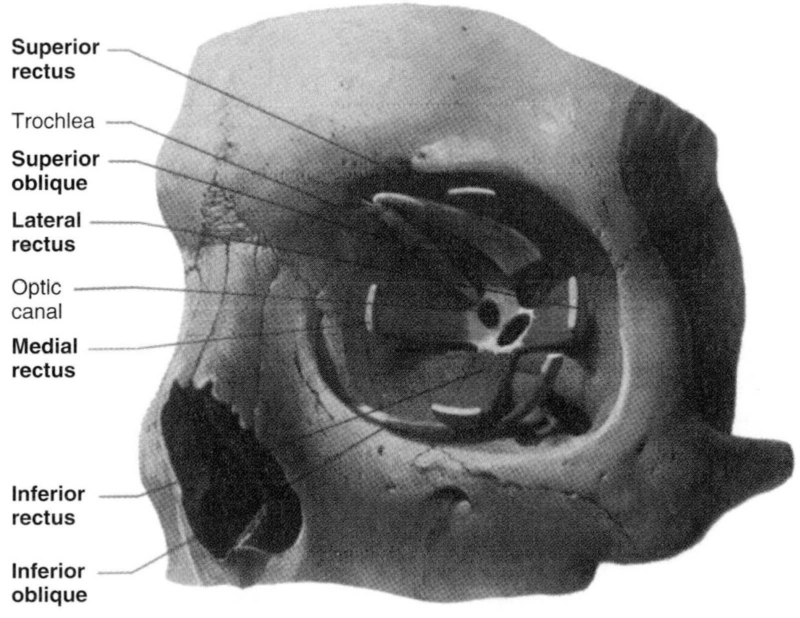

Figure 3-3
Extrinsic muscles of the left eye, anterior view

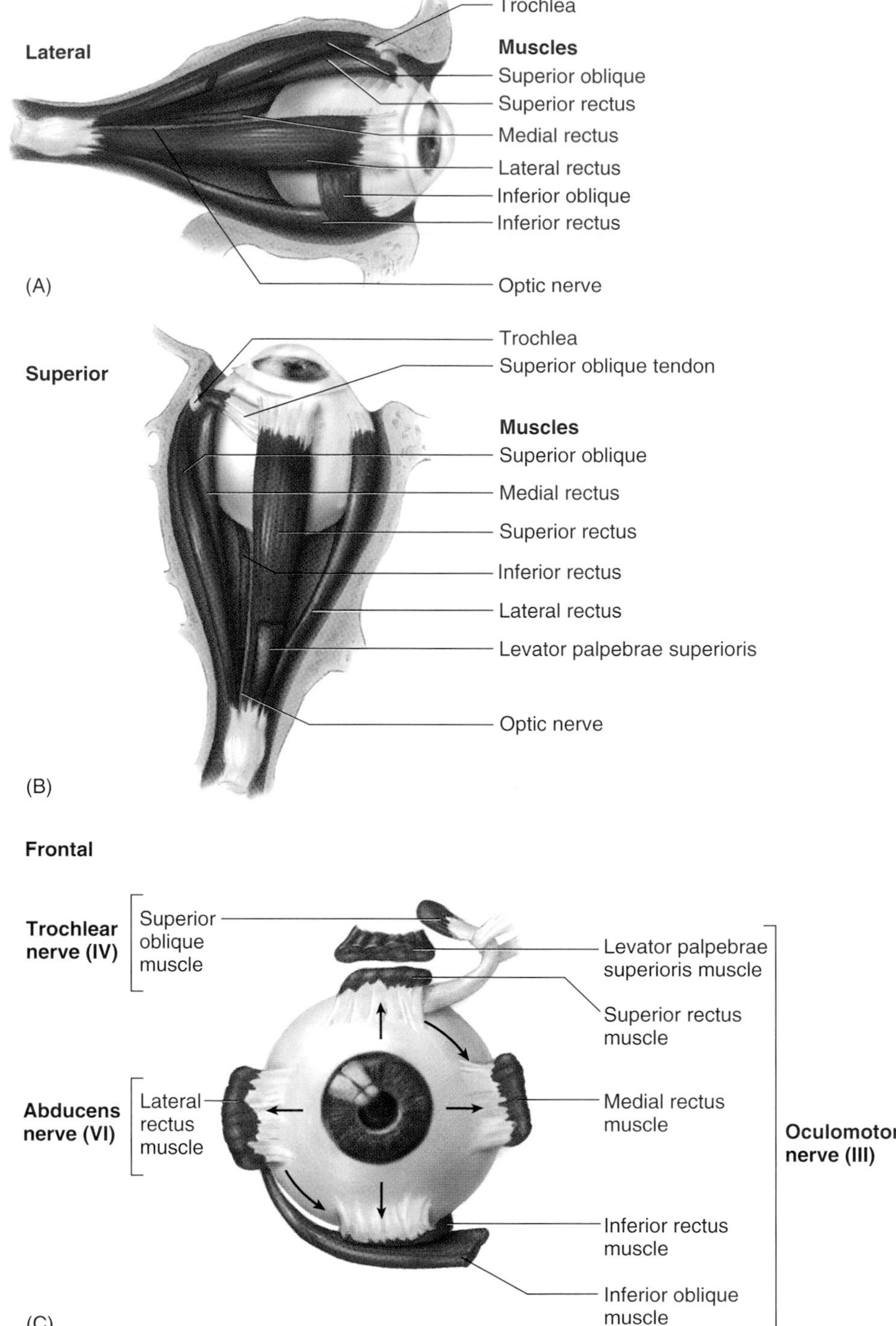

Figure 3-4
Extrinsic Muscles of the Eye. (a) Lateral view of the right eye. (b) Superior view of the right eye. (c) Innervation of the extrinsic muscles; *arrows* indicate the eye movement produced by each muscle.

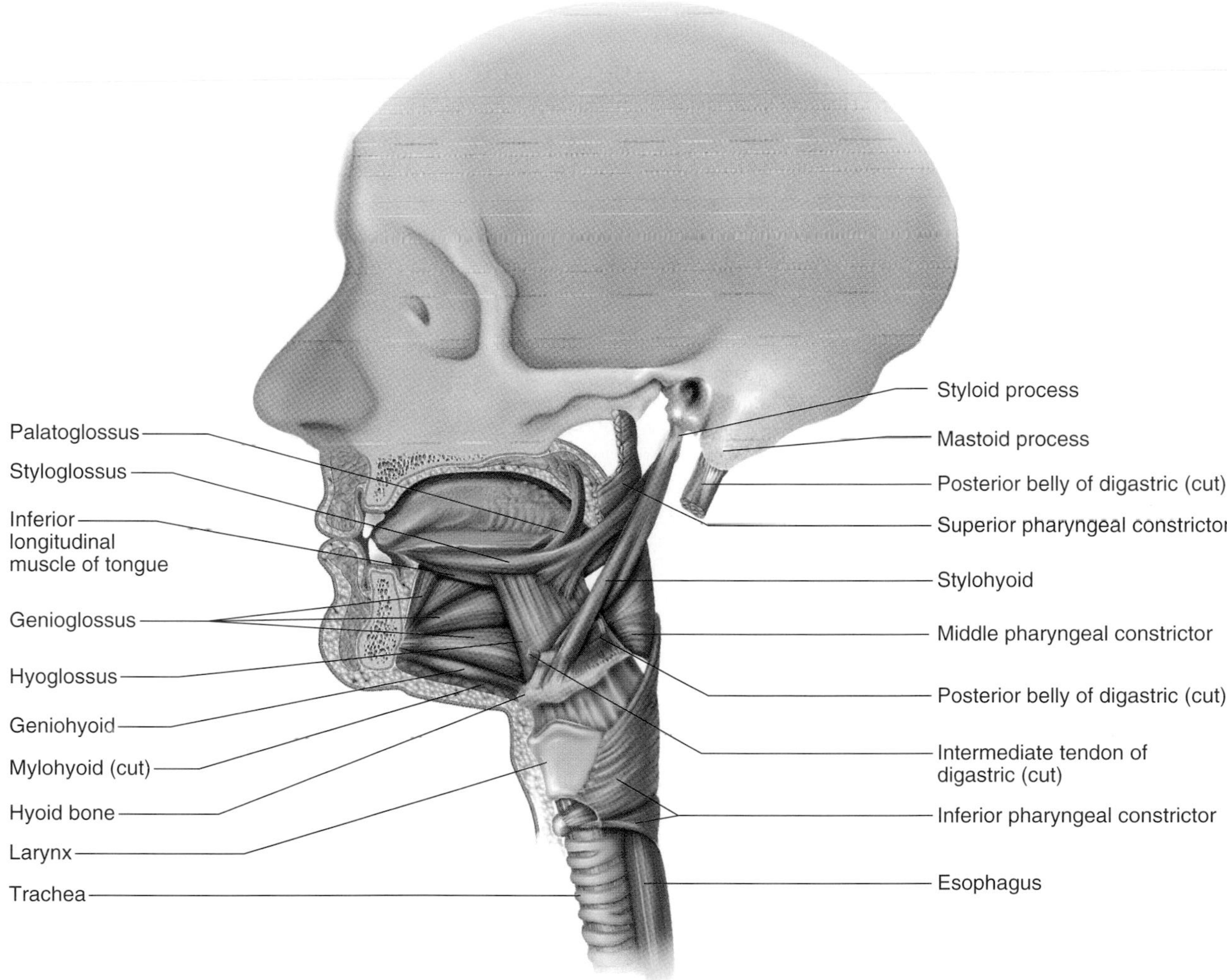

Figure 3-5
Muscles of the hypoglossal region and pharynx

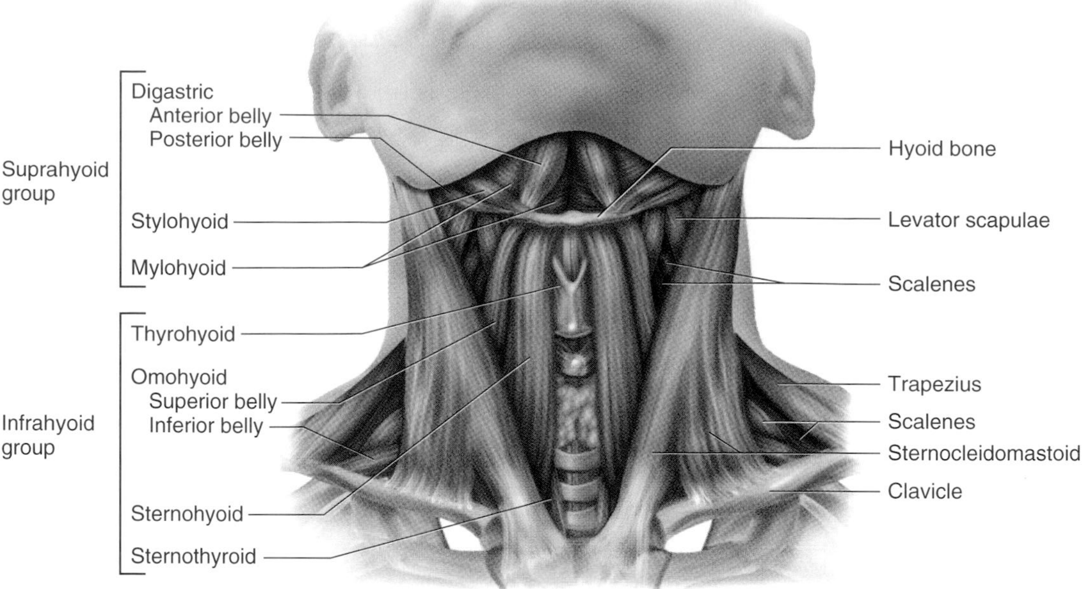

Figure 3-6
Muscles of the Neck, Anterior View

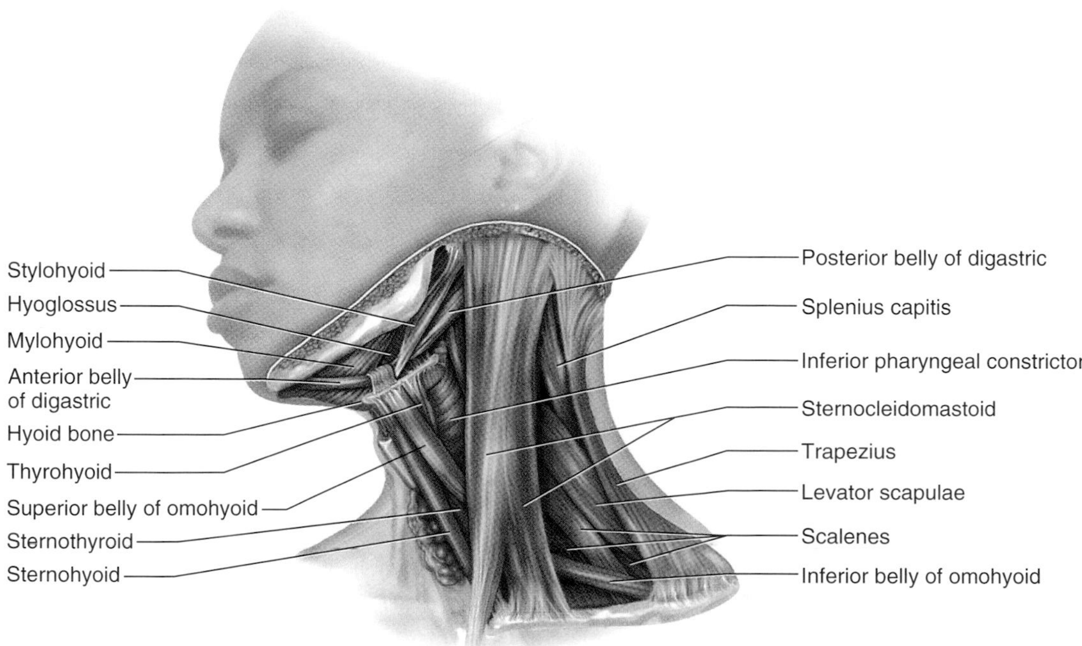

Figure 3-7
Muscles of the Neck, Left Lateral View

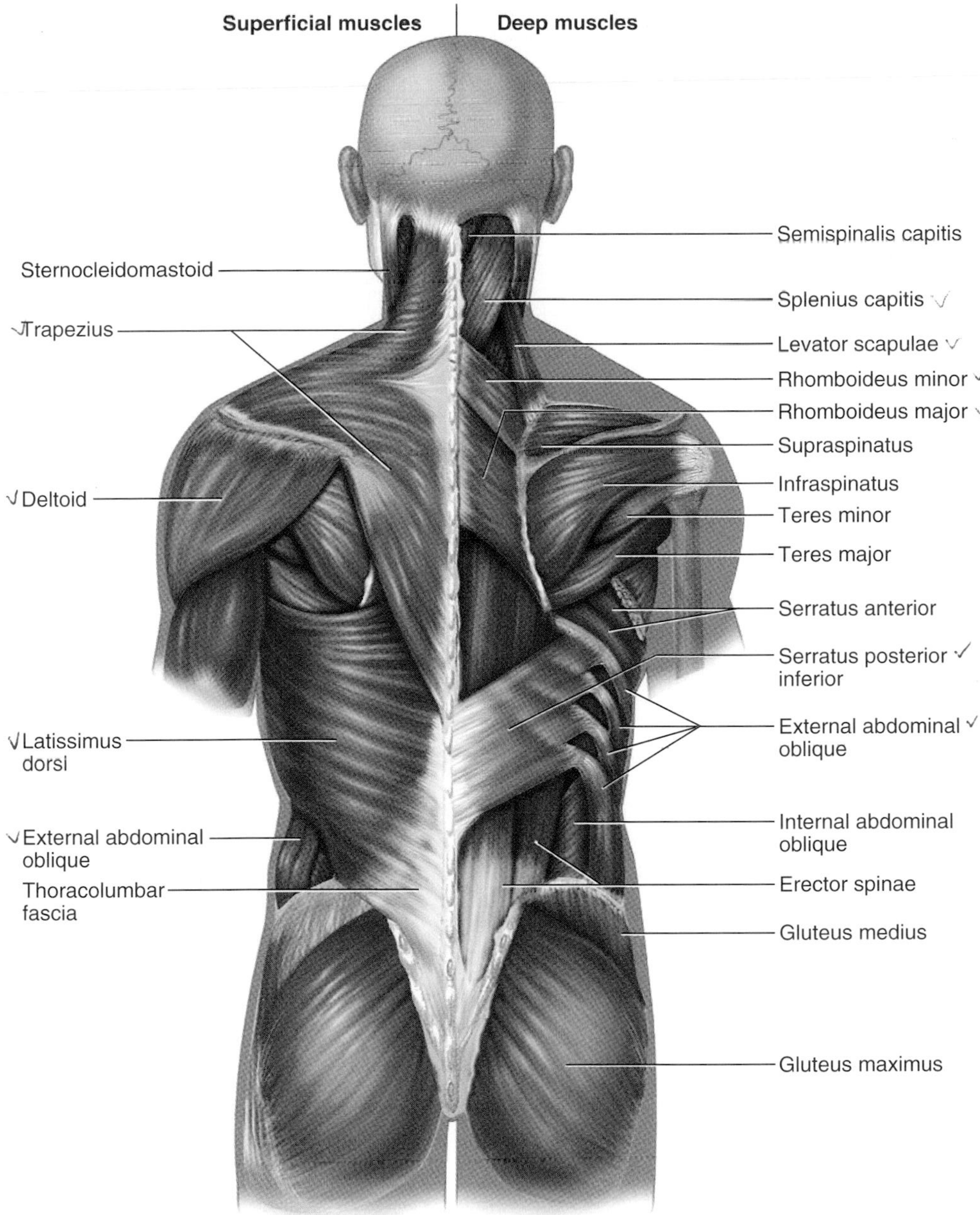

Figure 3-8

Neck, Back, and Gluteal Muscles. The most superficial muscles are shown on the left, and the next deeper layer on the right.

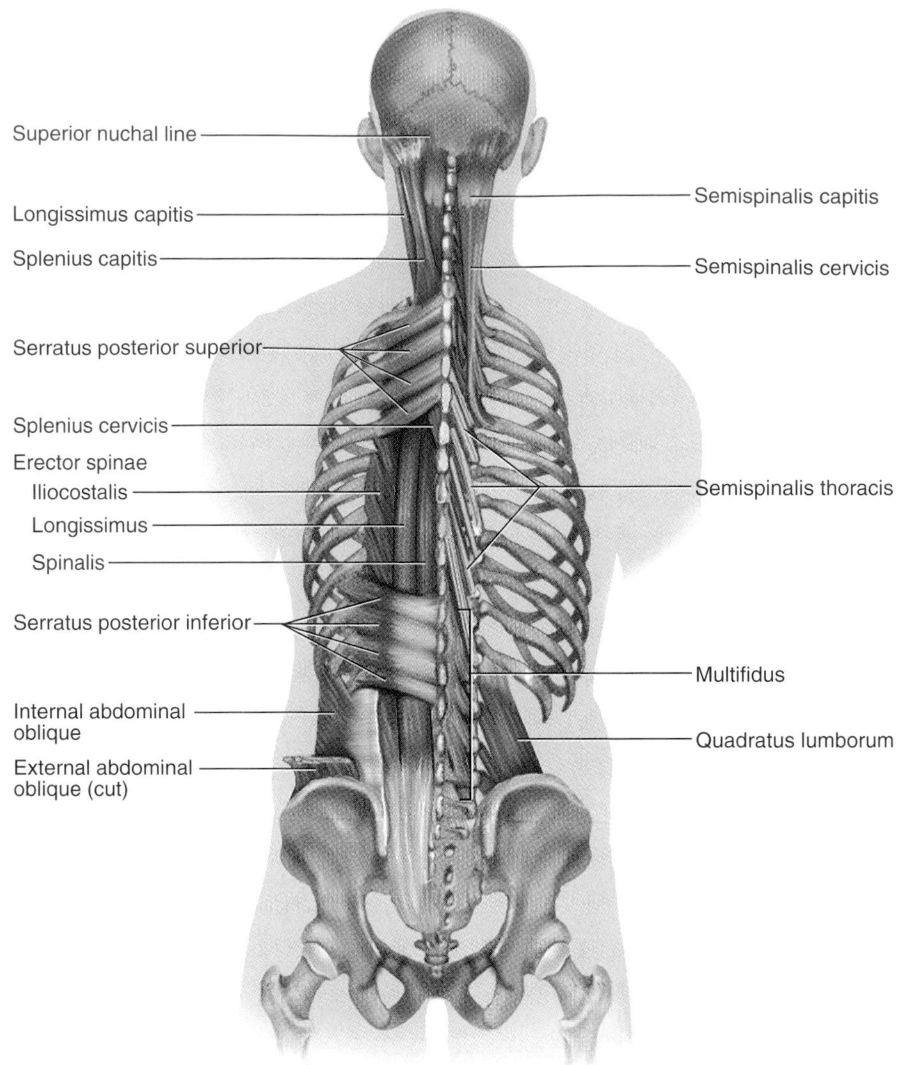

Figure 3-9

Muscles of the Back and Neck. Deep muscles of the back and the neck help move the head and hold the torso erect. The splenius capitis and semispinalis capitis are removed on the left to show underlying muscles. The iliocostalis, longissimus, and spinalis muscles compose the erector spinae muscle.

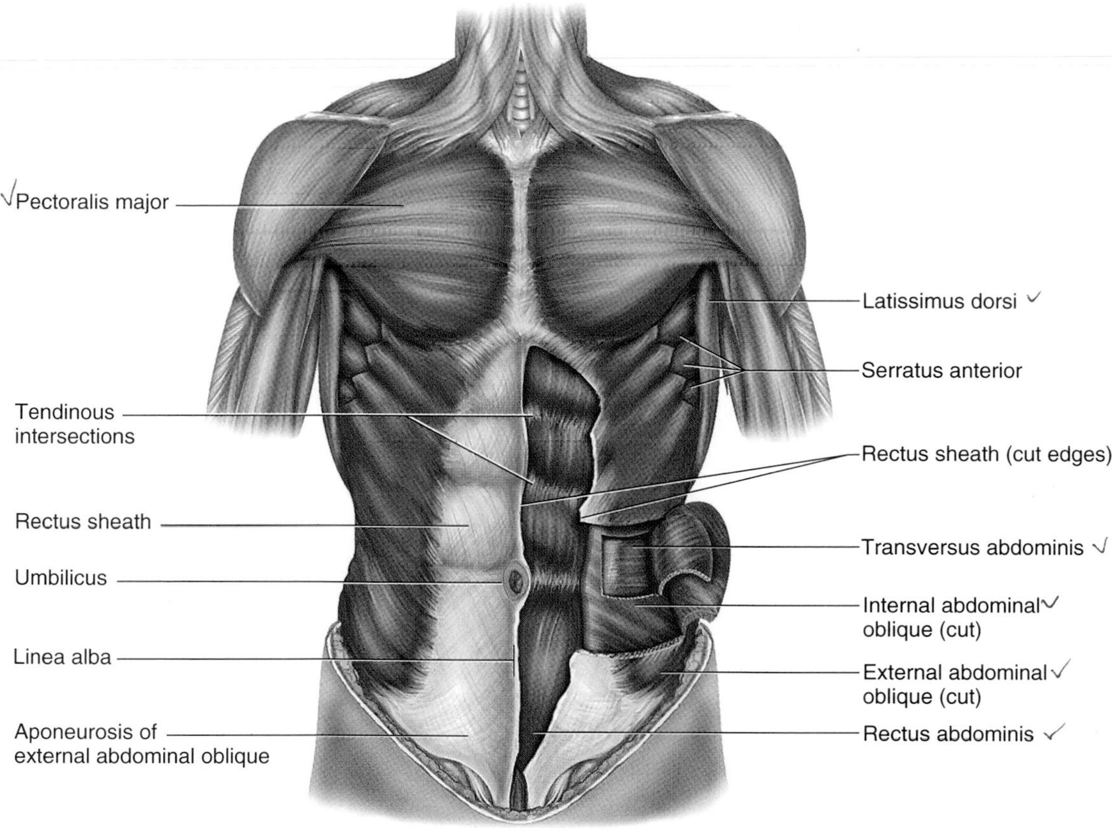

Figure 3-10

Thoracic and Abdominal Muscles. Superficial muscles. The left rectus sheath is cut away to expose the rectus abdominis muscle.

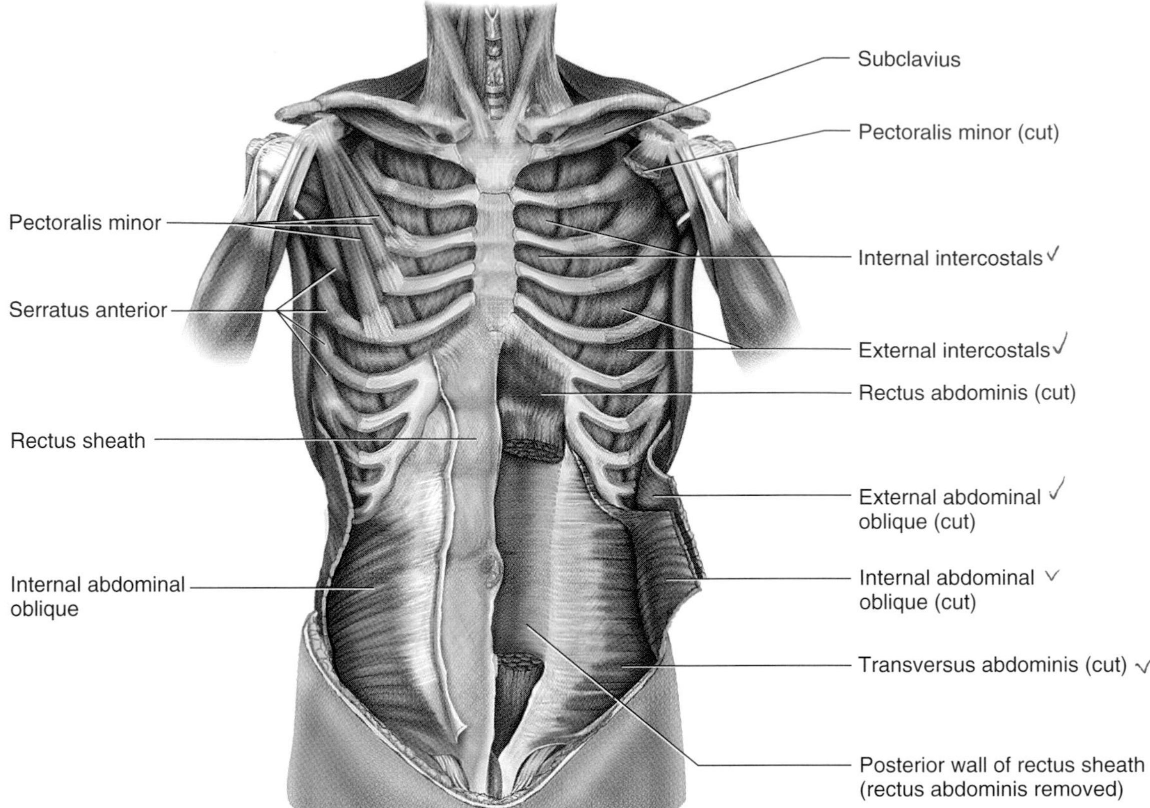

Figure 3-11
Thoracic and Abdominal Muscles. Deep muscles. On the anatomical right, the external abdominal oblique has been removed to expose the internal abdominal oblique and the pectoralis major has been removed to expose the pectoralis minor. On the anatomical left, the internal abdominal oblique has been cut to expose the transversus abdominis, and the rectus abdominis has been cut to expose the posterior rectus sheath.

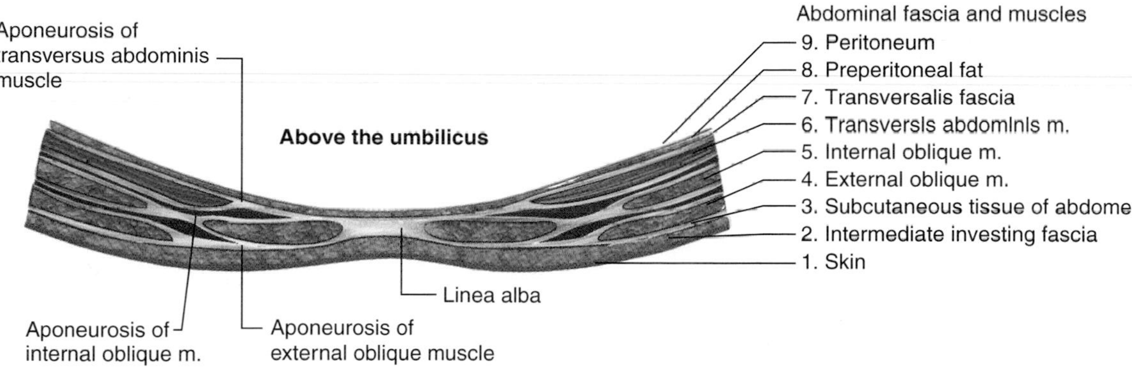

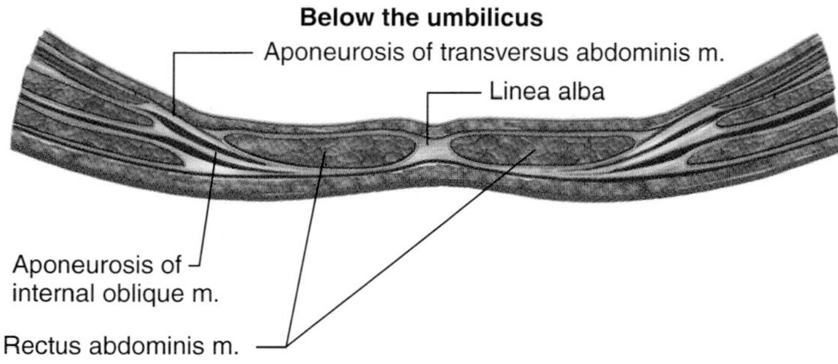

Figure 3-12
Muscles of the anterior abdominal wall, cross-sectional view above the umbilicus.

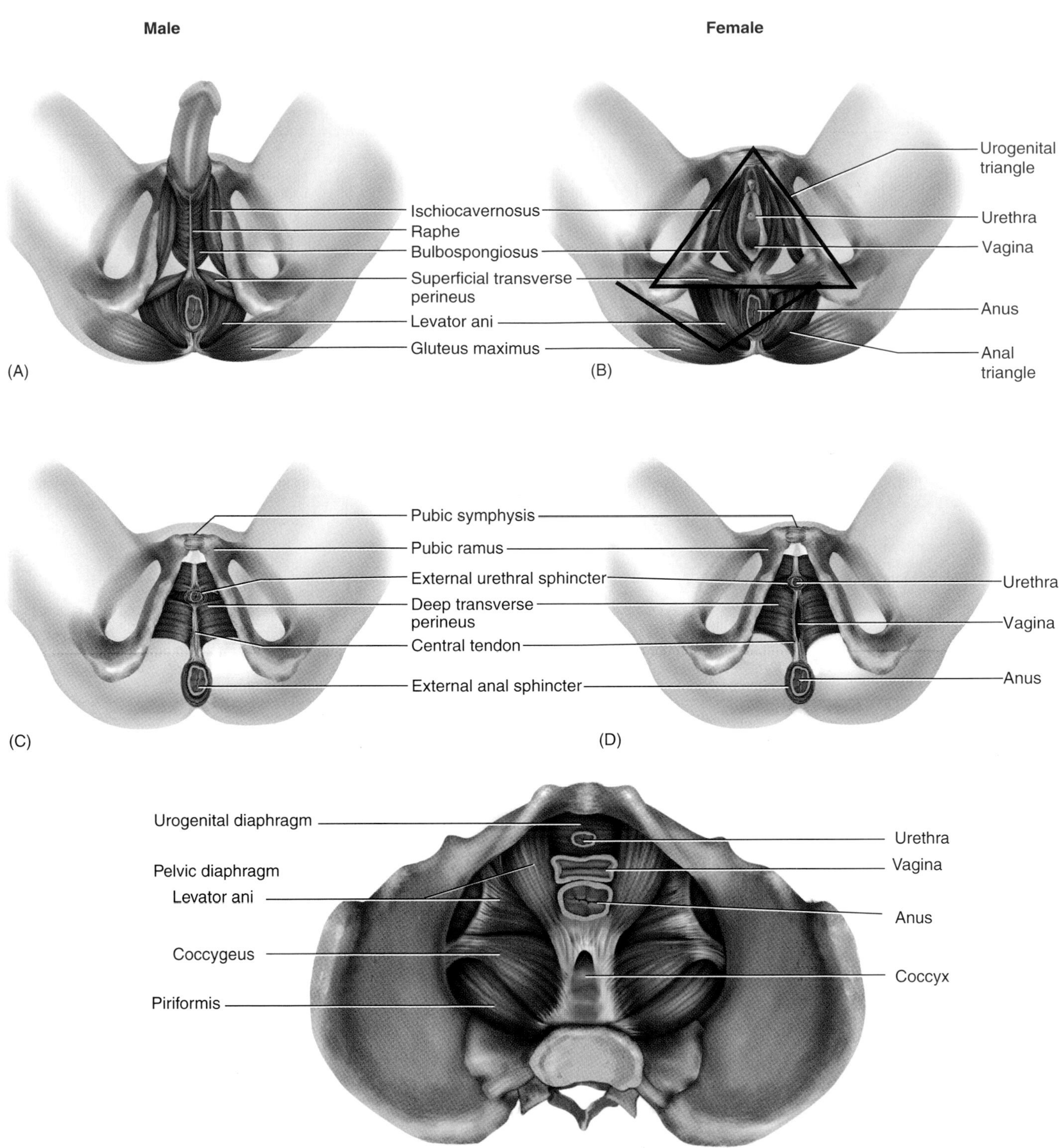

Figure 3-13
Muscles of the Pelvic Floor. (A, B) The superficial perineal space, inferior view. Triangles of the perineum are marked in b. (C, D) The urogenital diaphragm, inferior view; this is the next deeper layer after the muscles in a and b. (E) The pelvic diaphragm, the deepest layer, superior view (seen from within the pelvic cavity).

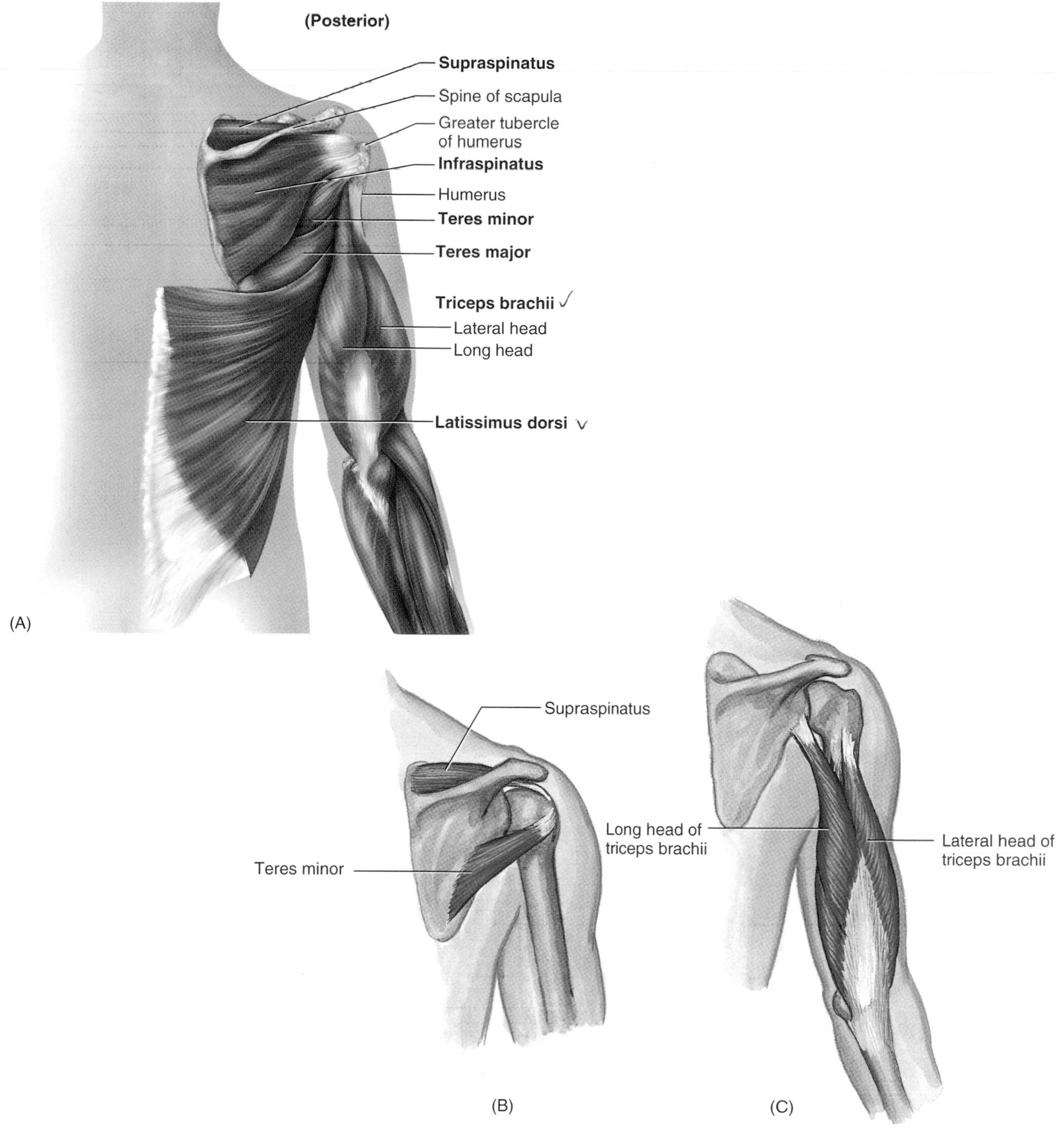

Figure 3-14
(A) Muscles of the posterior surface of the scapula and the arm; (B and C) muscles associated with the scapula.

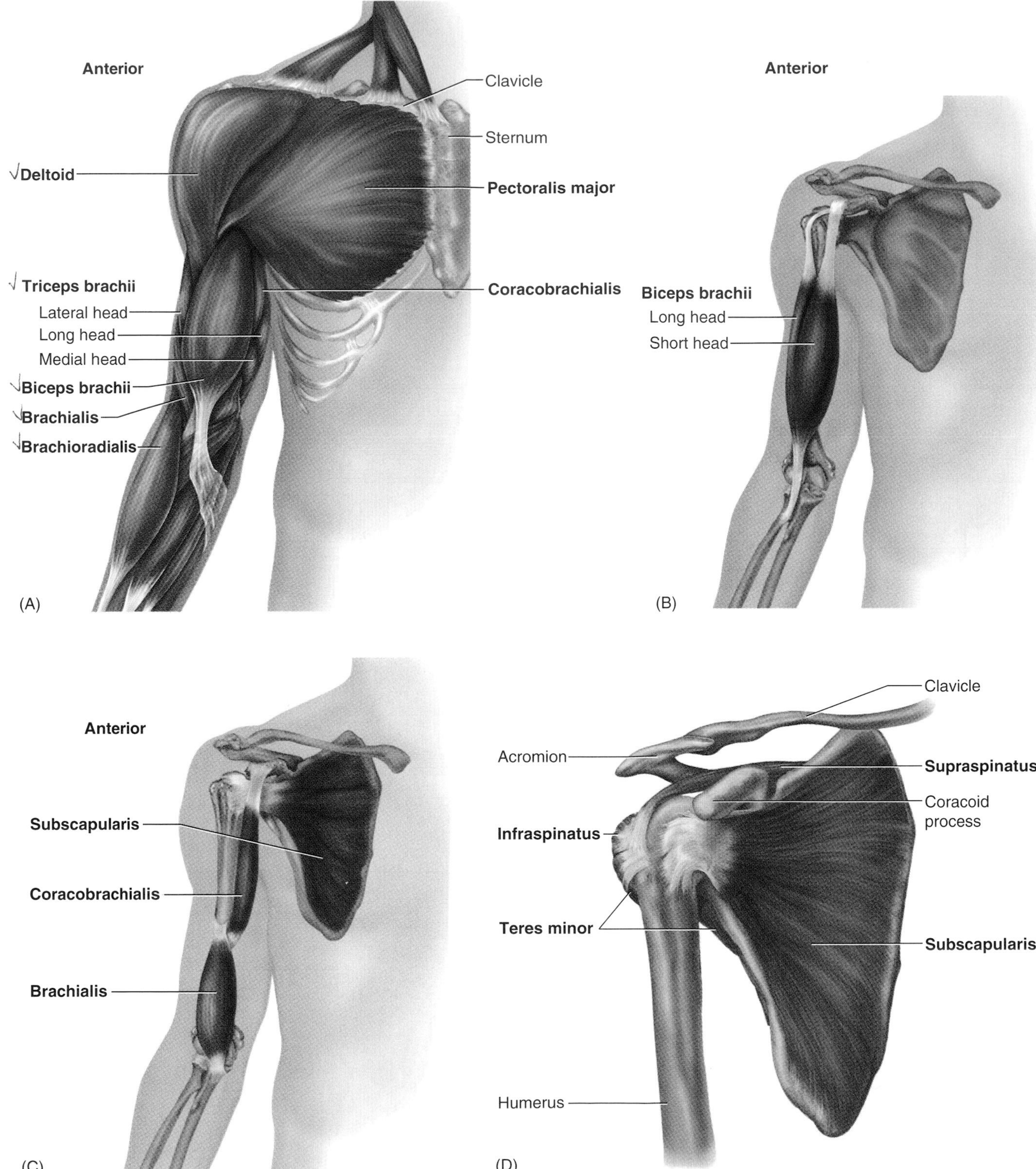

Figure 3-15
(A) Muscles of the anterior shoulder and the arm, with the rib cage removed. (B, B, and D) Isolated views of muscles associated with the arm.

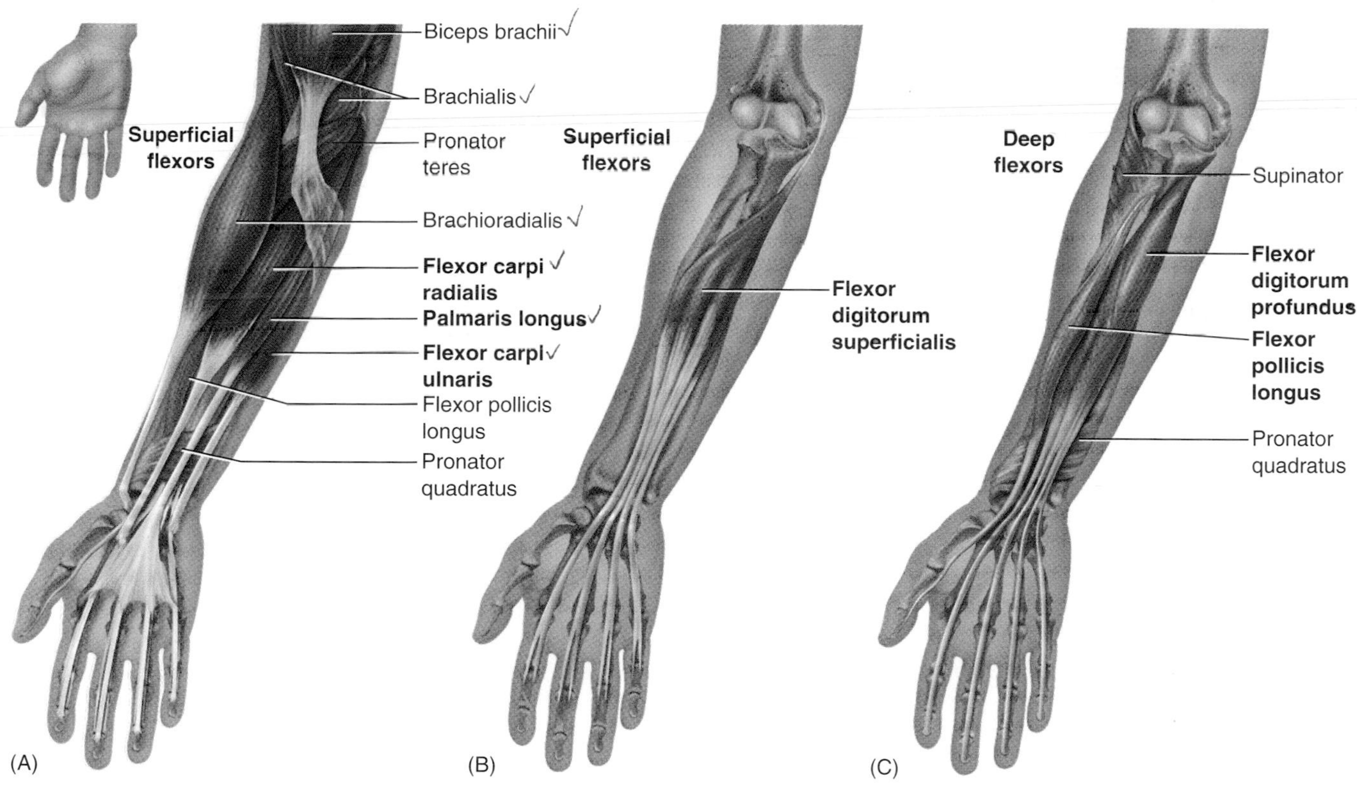

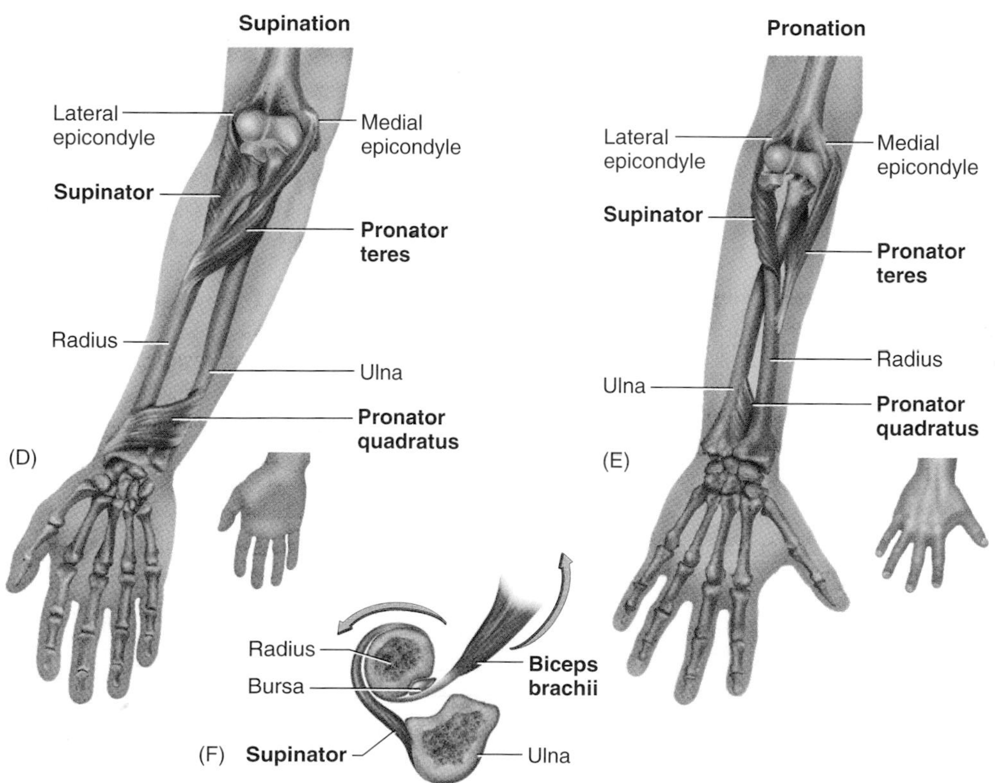

Figure 3-16

Muscles of the Forearm, Anterior View. (A) Superficial flexors; (B) the flexor digitorum superficialis, deep to the muscles in (A) but also classified as a superficial flexor; (C) deep flexors; (D) supinator, pronator teres, and pronator quadratus during supination; (E) during pronation; (F) during supination.

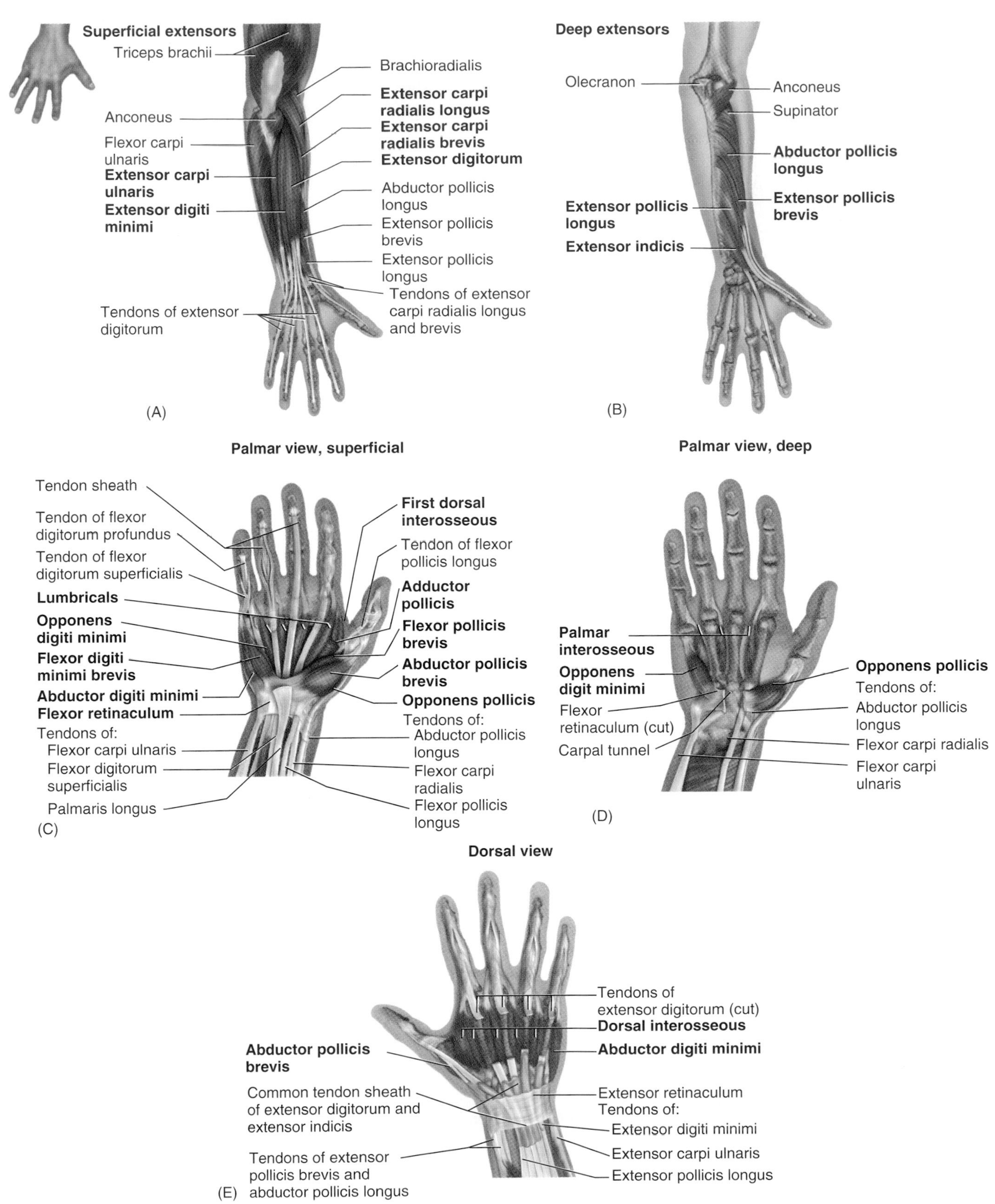

Figure 3-17
Muscles of the Forearm, Posterior View and Muscles of the Hand. (A) Forearm Superficial extensors; (B) Forearm deep extensors; (C) Superficial muscles and tendons of the hand, ventral view; (D) Deep muscles and tendons of the hand, ventral view; (E) Major extensor muscles and tendons of the hand, dorsal view.

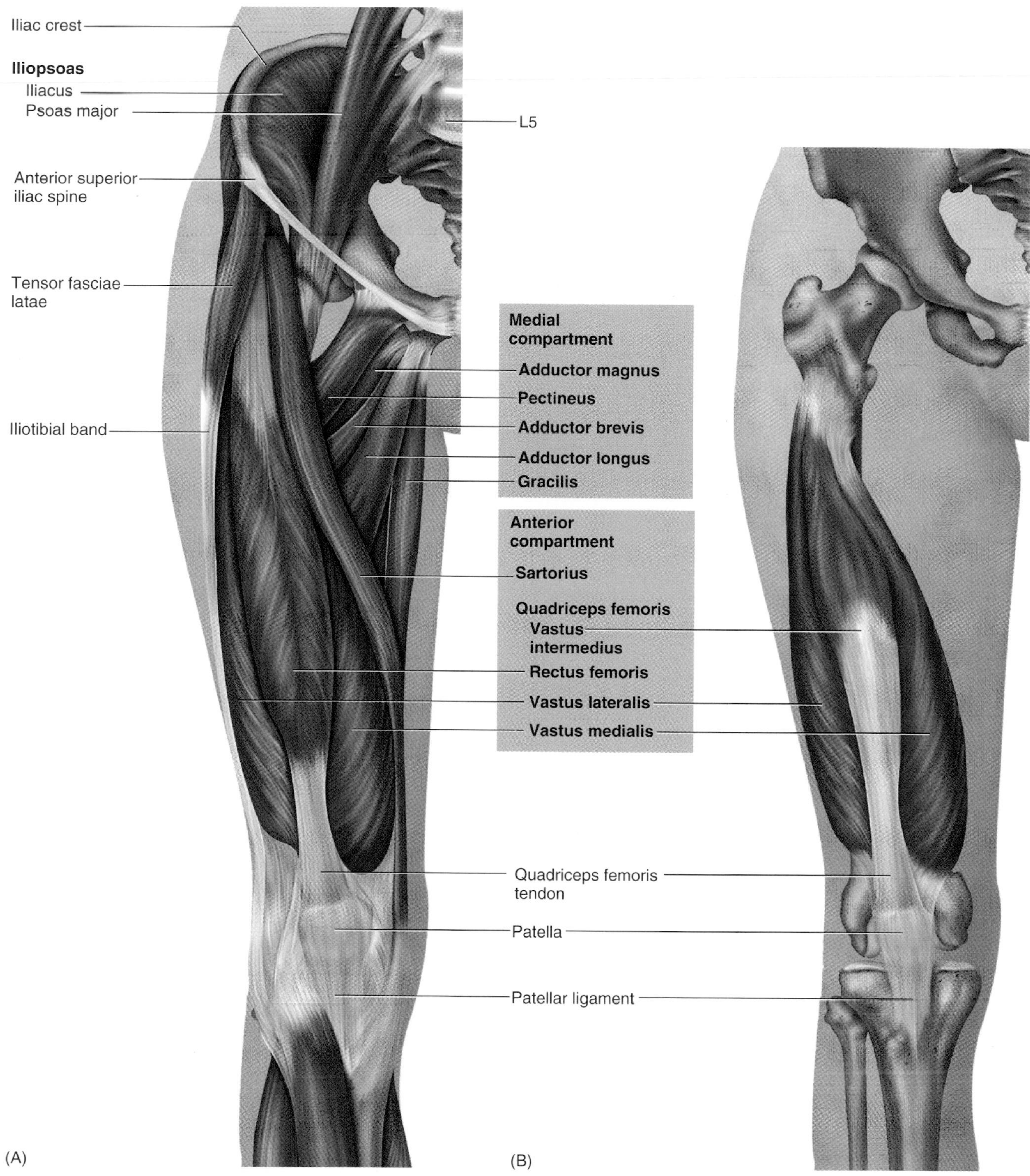

Figure 3-18
Muscles of the Thigh, Anterior View. (A) Superficial muscles; (B) rectus femoris and other muscles removed to expose the other three heads of the quadriceps femoris.

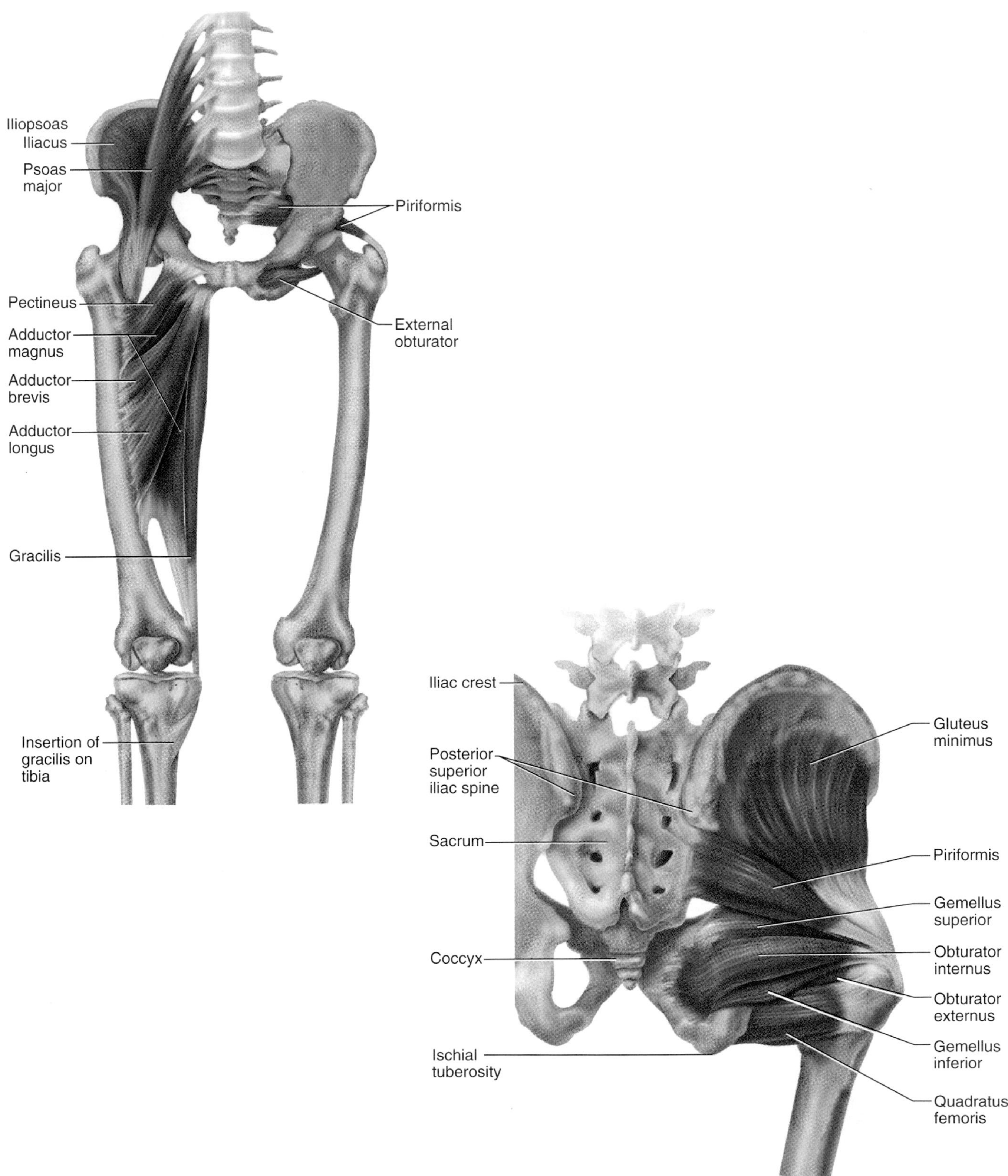

Figure 3-19
(A) Major adductors of the thigh, ventral view, (B) Deeper muscles of the hip, dorsal view

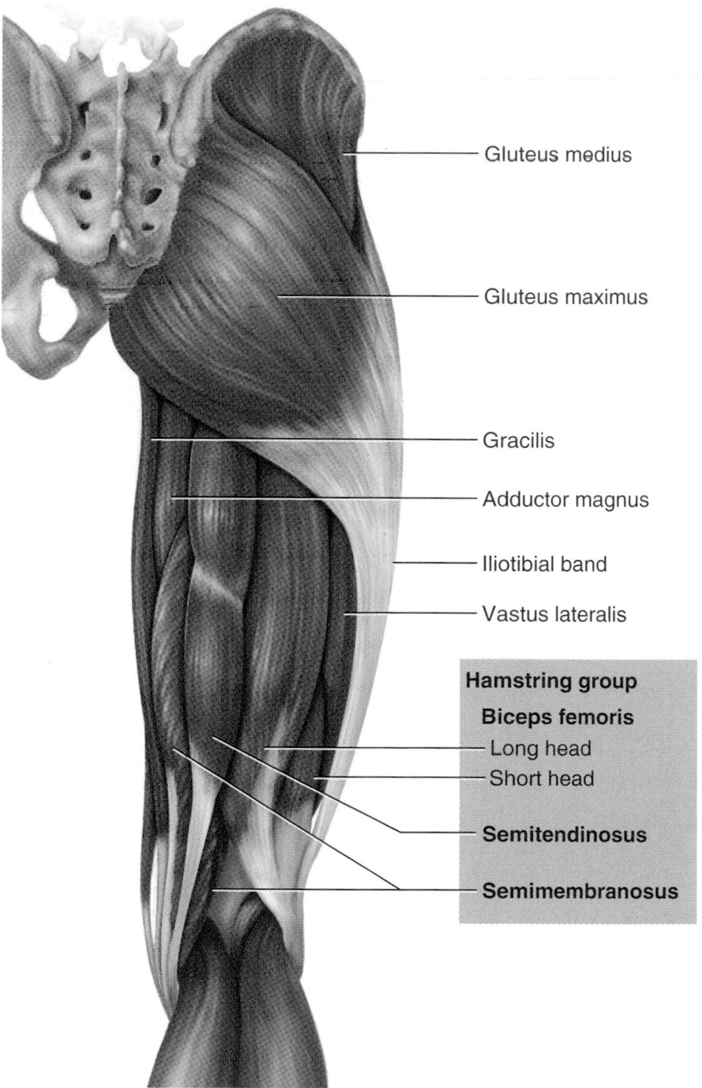

Figure 3-20
Gluteal and Thigh Muscles, Posterior View

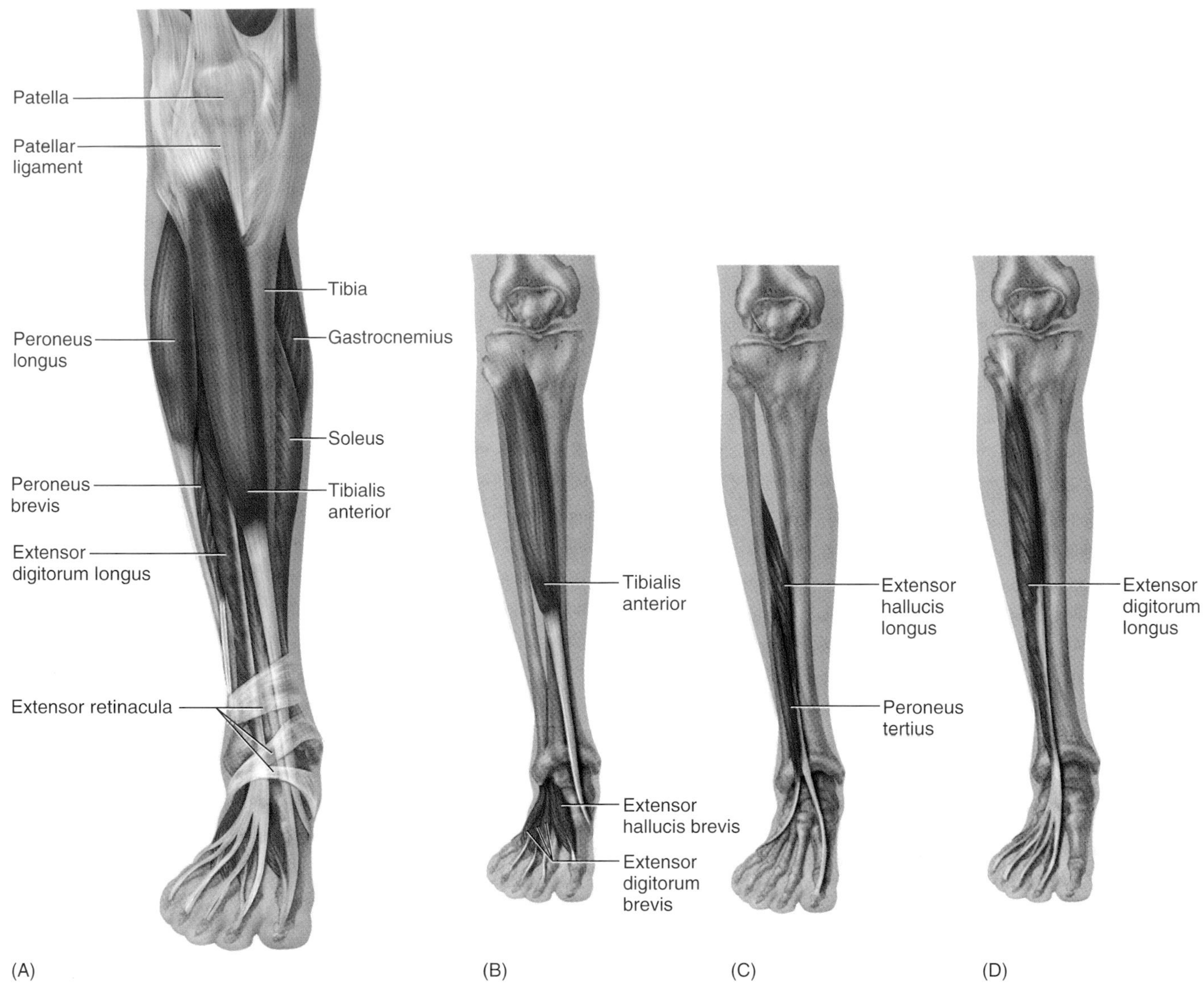

Figure 3-21
(A) Muscles of the anterior right leg. (B–D) Isolated views of muscles associated with the anterior leg.

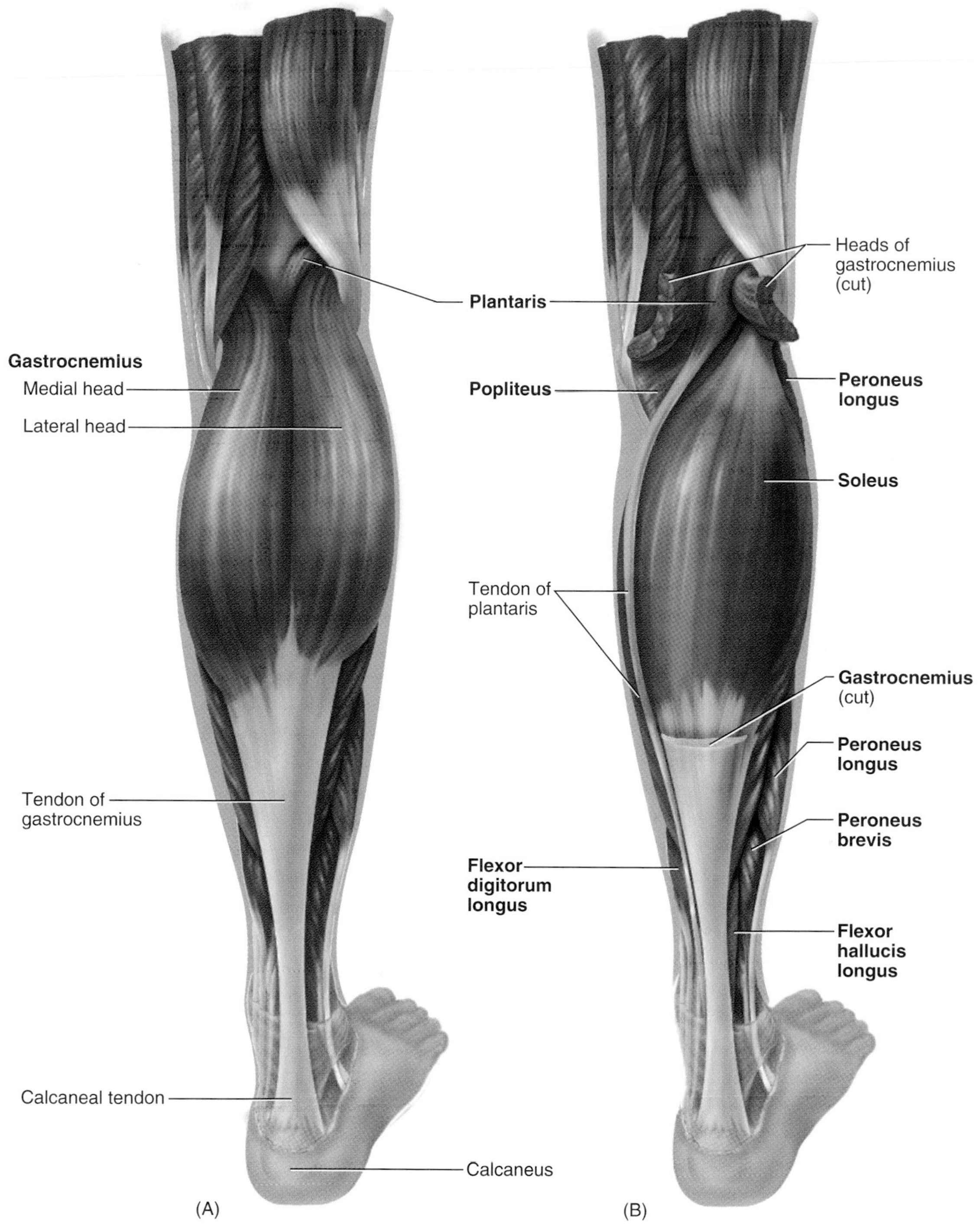

Figure 3-22
(A) Superficial muscles of the lower right leg. (B) Deeper muscles of the lower right leg.

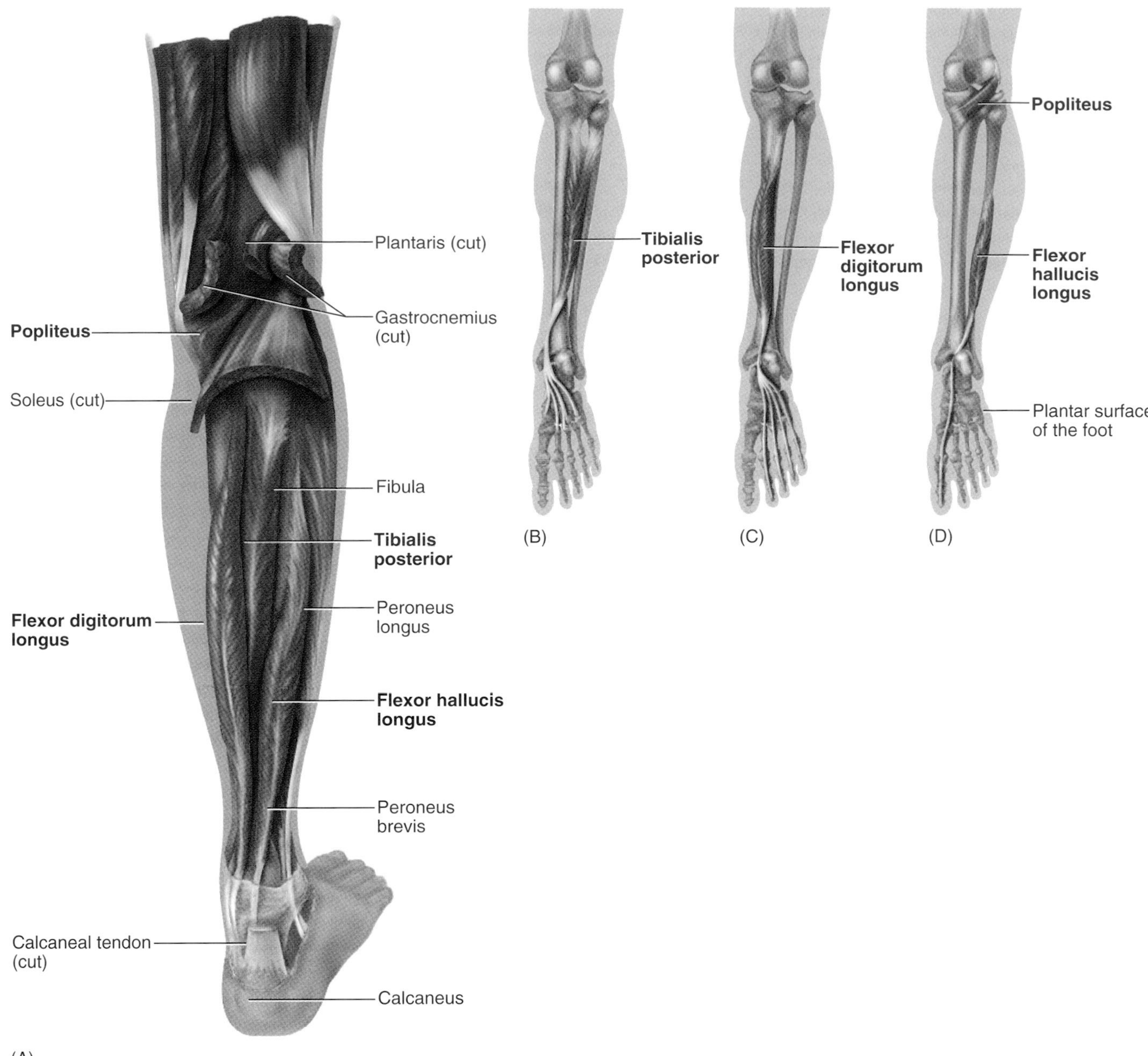

Figure 3-23
(A) Muscles of the posterior right leg. (B–D) Isolated views of muscles associated with the posterior right leg.

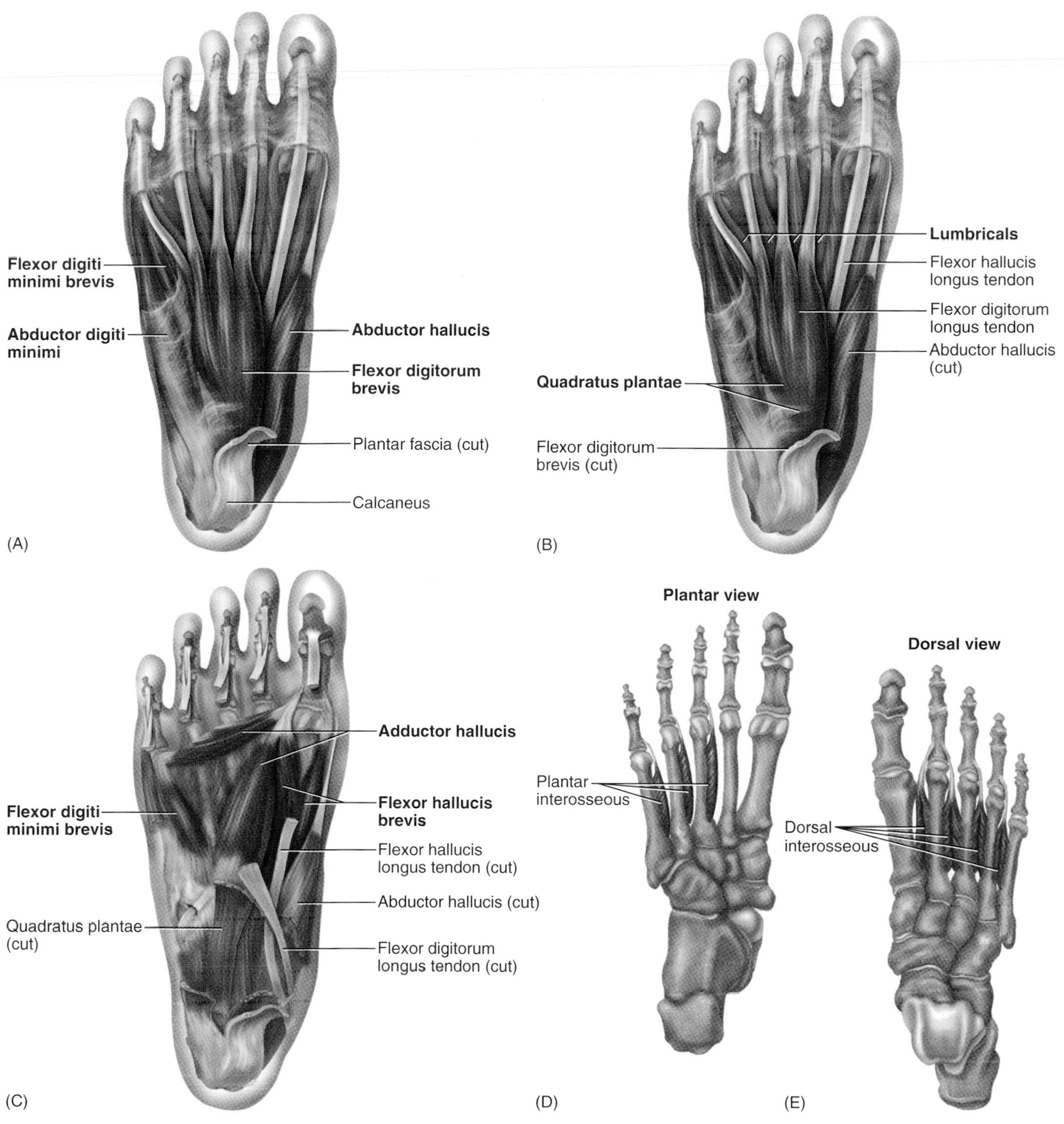

Figure 3-24
Intrinsic Muscles of the Foot. (A–D) First through fourth layers, respectively, in ventral (plantar) views; (E) fourth layer, dorsal view.

CHAPTER 4

Dissections

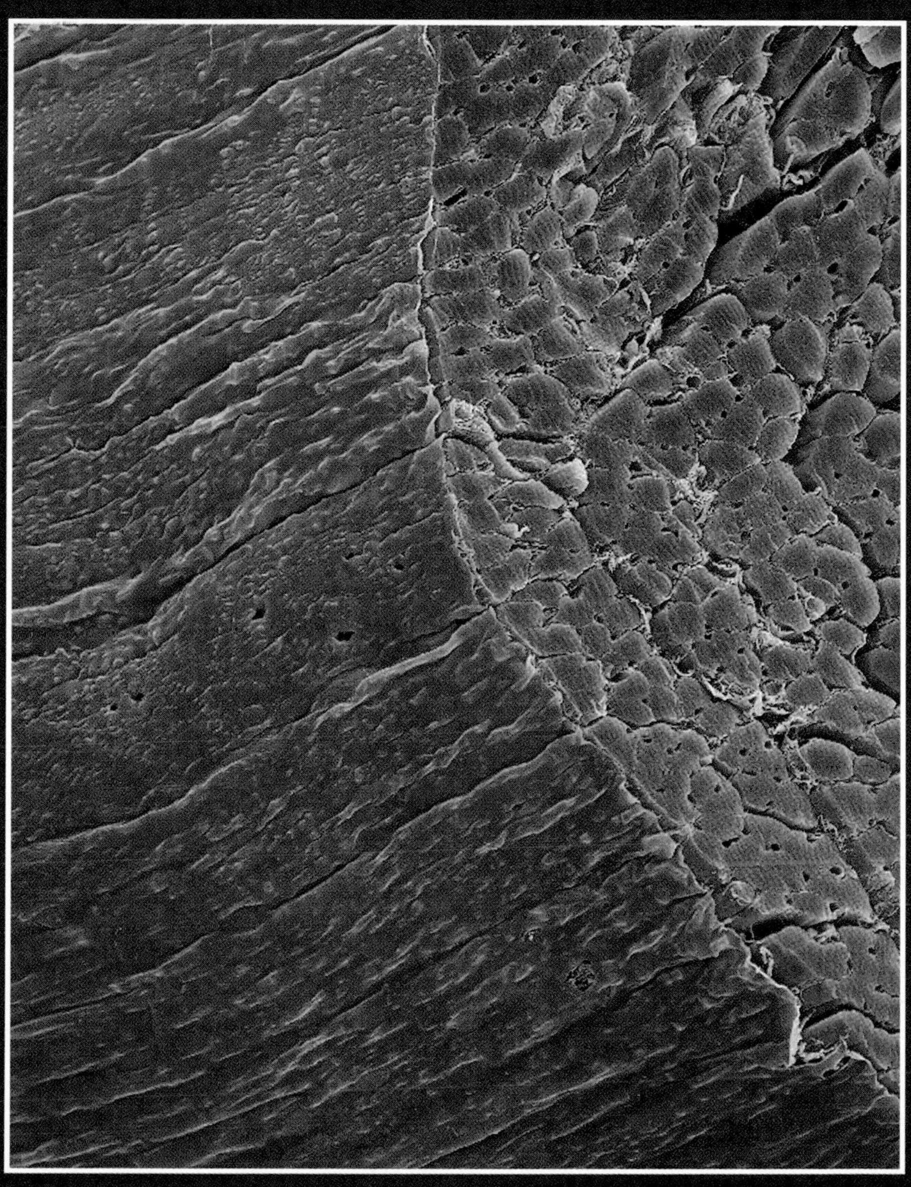

Scanning electron micrograph (SEM) of a transverse section through skeletal muscle

Figure 4-1
Superficial Anatomy of Cat Head and Neck, Left Lateral View

1. Vibrissal barrels for sensory hairs (whiskers)
2. Tongue
3. Buccinator muscle
4. Diagastric muscle
5. Temporalis muscle
6. Masseter muscle
7. Dorsal buccal branch of facial (VII) nerve
8. Ventral buccal branch of facial (VII) nerve
9. Parotid duct
10. Parotid gland
11. Submandibular gland
12. Lymph node
13. External jugular vein
14. Transverse jugular vein
15. Anterior facial vein
16. Posterior facial vein
17. Sternohyoid muscle
18. Sternothyroid muscle
19. Cleidomastoid muscle
20. Sternomastoid muscle
21. Clavotrapezius muscle
22. Clavobrachialis muscle
23. Acromiotrapezius muscle

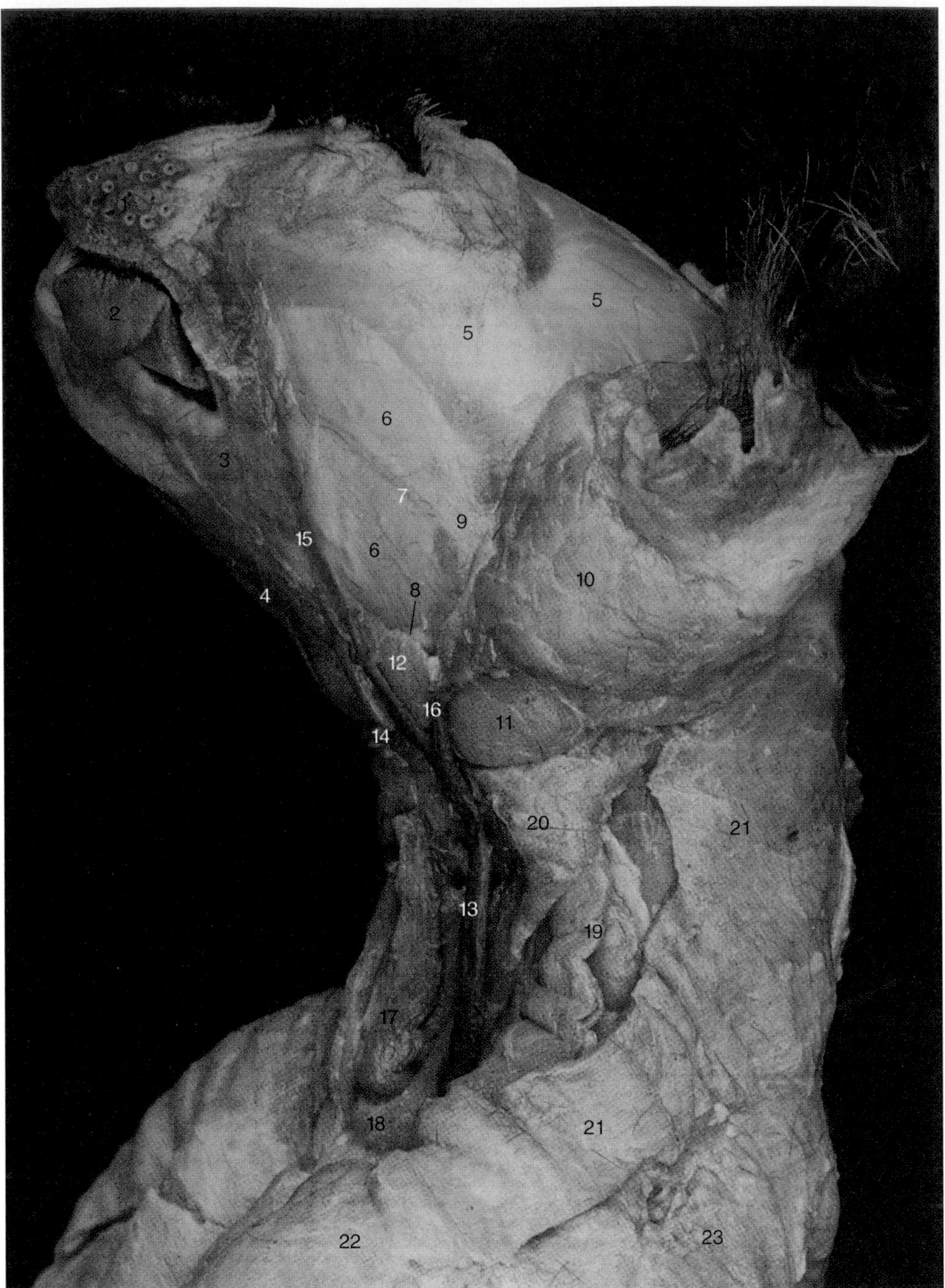

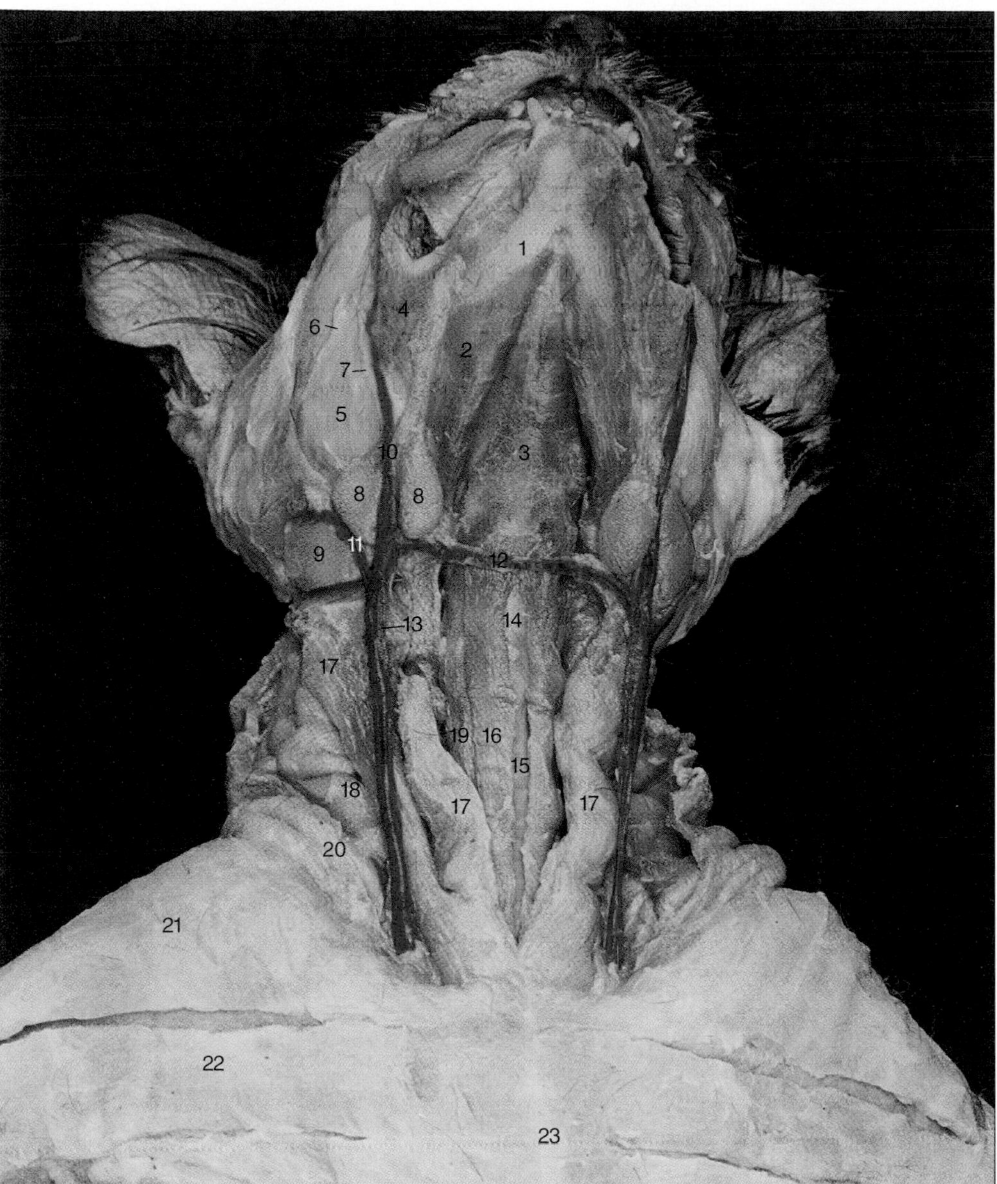

Figure 4-2
Superficial Anatomy of Cat Head and Neck, Ventral View

1. Body of mandible
2. Digastric muscle
3. Mylohyoid muscle
4. Buccinator muscle
5. Masseter muscle
6. Dorsal branch of facial (VII) nerve
7. Ventral branch of facial (VII) nerve
8. Lymph node
9. Submandibular gland
10. Anterior facial vein
11. Posterior facial vein
12. Transverse jugular vein
13. External jugular vein
14. Larynx
15. Trachea
16. Sternohyoid muscle
17. Sternomastoid muscle (unavoidably damaged on animal's right side during vascular perfusion)
18. Cleidomastoid muscle
19. Sternothyroid muscle
20. Clavotrapezius muscle
21. Clavobrachialis muscle
22. Pectoantebrachialis muscle
23. Sternum

Figure 4-3
Deep Anatomy of Cat Head and Neck, Left Ventrolateral View

1. Lower canine tooth
2. Upper canine tooth
3. Upper premolar tooth
4. Lower premolar tooth
5. Body of mandible
6. Digastric muscle
7. Mylohyoid muscle
8. Temporalis muscle
9. Masseter muscle
10. Dorsal branch of facial (VII) nerve
11. Ventral branch of facial (VII) nerve
12. Parotid duct
13. Cutaneous branch of facial (VII) nerve
14. Platysma muscle (reflected)
15. Lymph node
16. Sternohyoid muscle
17. Sternomastoid muscle (reflected)
18. Cleidomastoid muscle
19. Omohyoid muscle
20. 4th cervical nerve
21. 5th cervical nerve
22. External jugular vein
23. Subclavian vein
24. Musculocutaneous nerve
25. Radial nerve
26. Median nerve
27. Ulnar nerve
28. Thoracic nerve
29. Ventral thoracic nerve (cut)
30. Axillary nerve

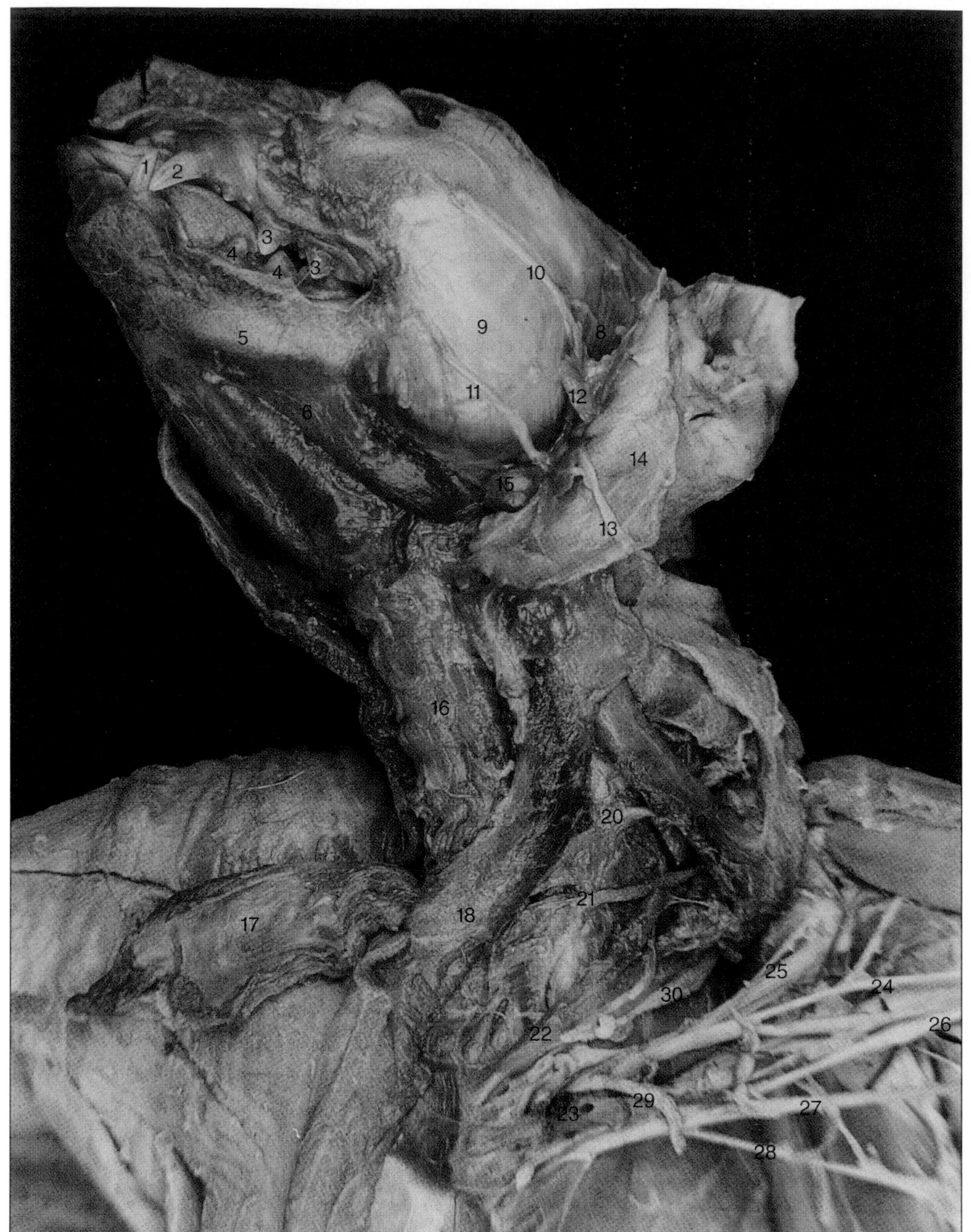

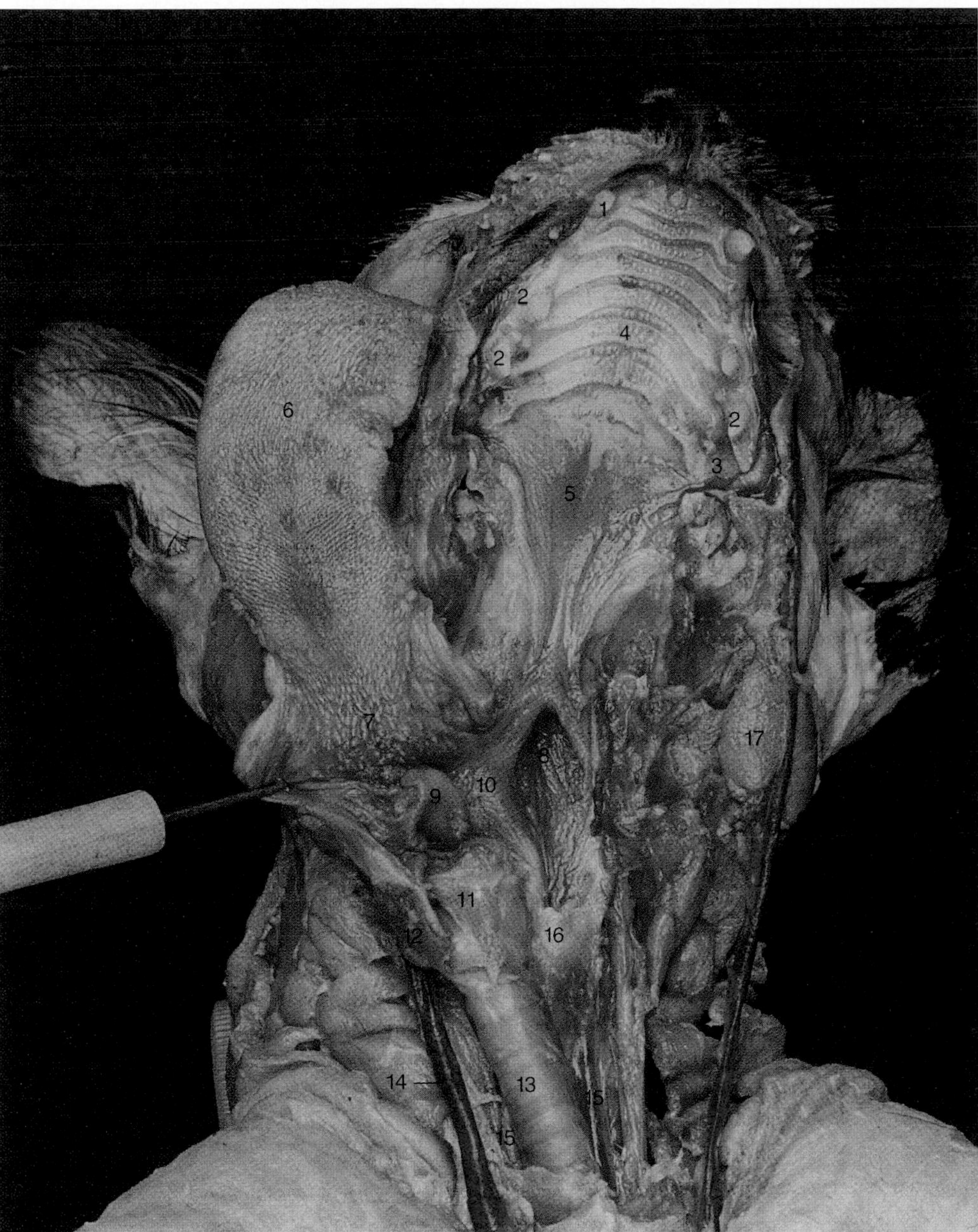

Figure 4-4
Deep Anatomy of Cat Head and Neck, Oral Cavity with Mandlble Removed

1. Upper canine tooth
2. Upper premolar tooth
3. Upper molar tooth
4. Hard palate with palatine rugae
5. Soft palate
6. Tongue
7. Lingual tonsils
8. Isthmus of fauces
9. Epiglottis
10. Palatine tonsil
11. Larynx
12. Thyroid gland (reflected)
13. Trachea
14. External jugular vein
15. Common carotid artery
16. Esophagus
17. Lymph node

Figure 4-5
Superficial Muscles of the Cat Thoracic Limb, Ventral View

1. Clavobrachialis muscle
2. Pectoantebrachialis muscle
3. Pectoralis major muscle
4. Pectoralis minor muscle
5. Latissimus dorsi muscle
6. Epitrochlearis muscle
7. Flexor carpi ulnaris muscle
8. Palmaris longus muscle
9. Flexor carpi radialis muscle
10. Pronator teres muscle
11. Extensor carpi radialis muscle
12. Brachioradialis muscle (cut)
13. Antebrachial fascia
14. Olecranon process of ulna
15. Flexor retinaculum (transverse carpal ligament)

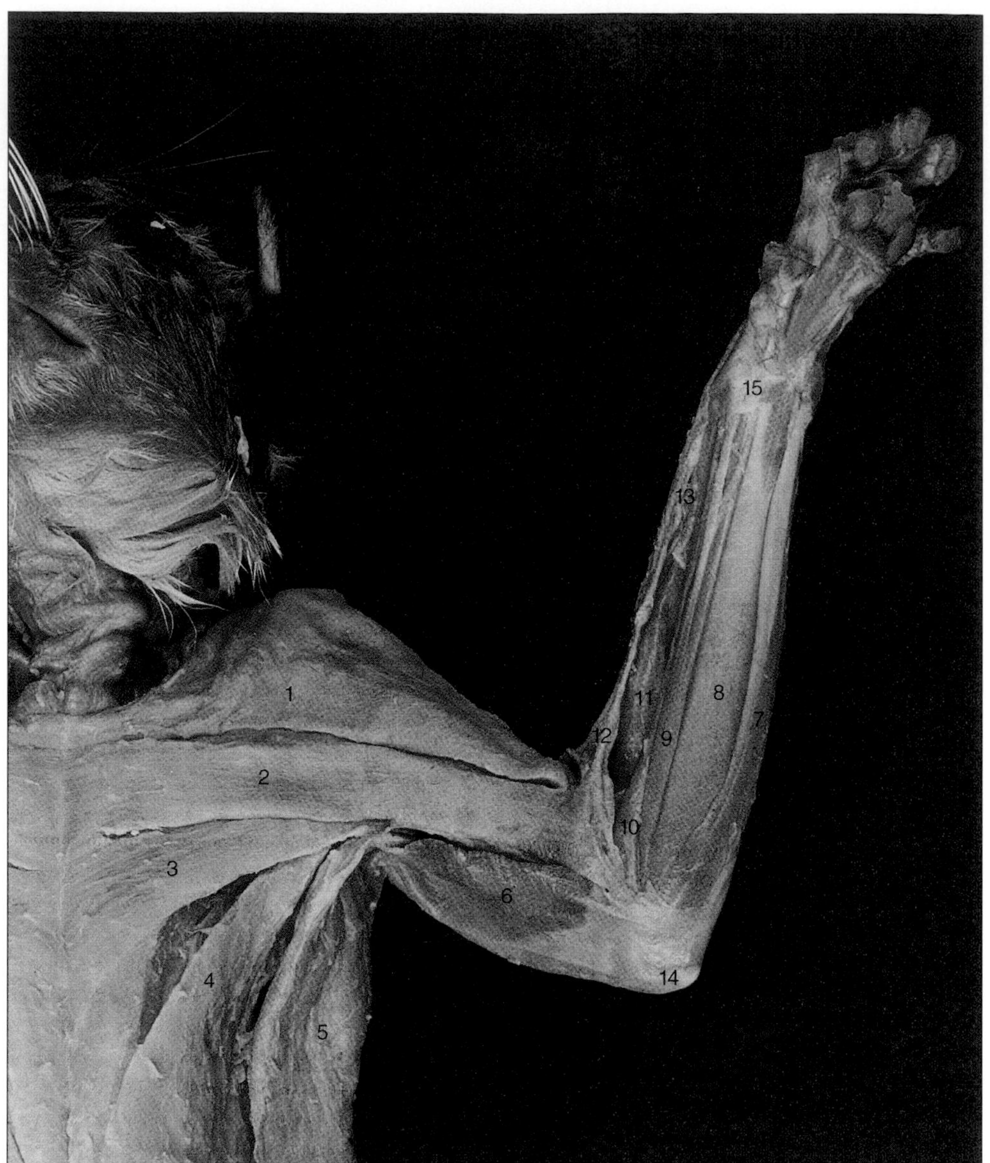

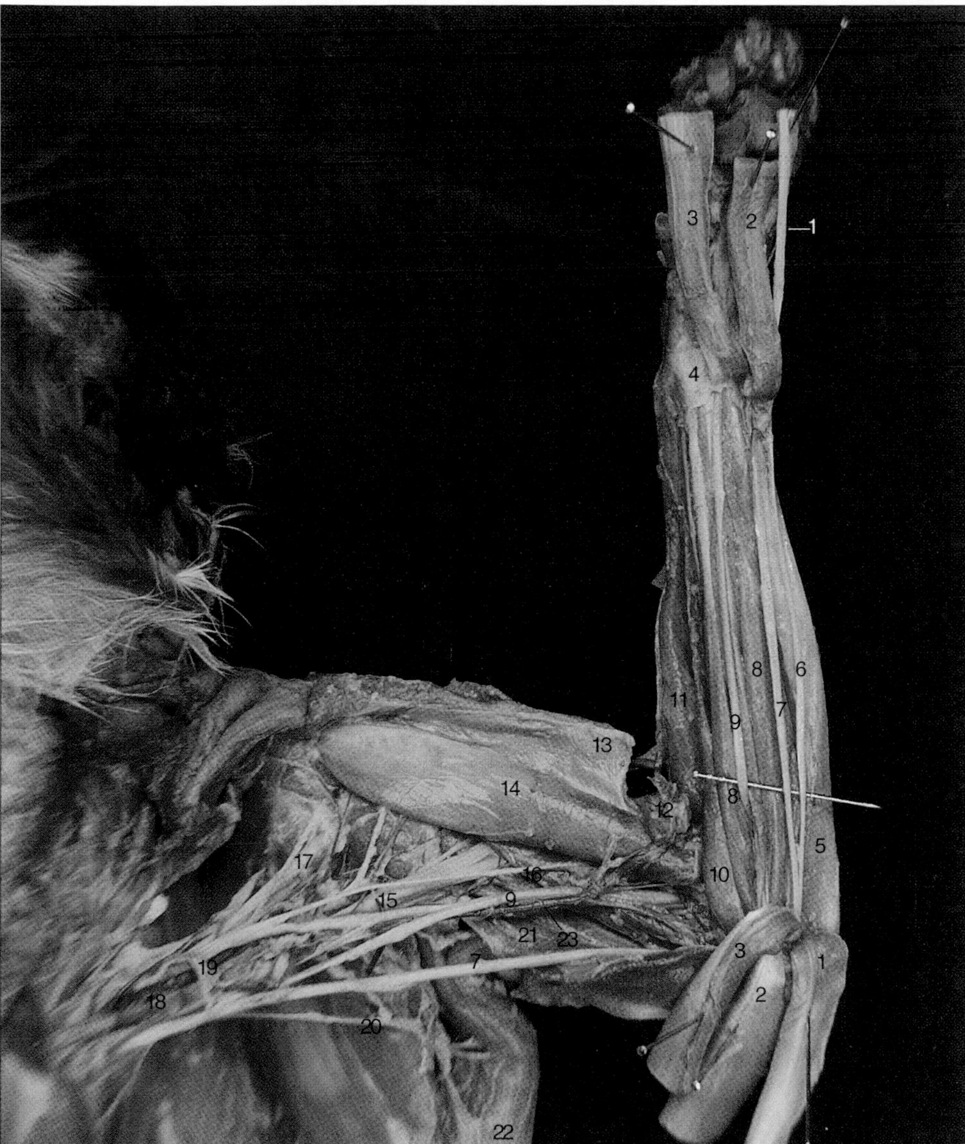

Figure 4-6
Deep Muscles of the Cat Left Thoracic Limb, Ventral View

1. Flexor carpi ulnaris muscle (cut and reflected)
2. Palmaris longus muscle (cut and reflected)
3. Flexor carpi radialis muscle (cut and reflected)
4. Flexor retinaculum
5. Extensor carpi ulnaris
6. Cutaneous branch of ulnar nerve
7. Ulnar nerve
8. Flexor digitorum profundus
9. Median nerve
10. Pronator teres muscle
11. Extensor carpi radialis muscle
12. Brachioradialis muscle (cut)
13. Clavobrachialis muscle (cut and reflected)
14. Biceps brachii muscle
15. Radial nerve
16. Musculocutaneous nerve
17. Axillary nerve
18. Subclavian vein
19. Ventral thoracic nerve (cut)
20. Thoracic nerve
21. Triceps brachii muscle
22. Latissimus dorsi muscle
23. Brachial vein

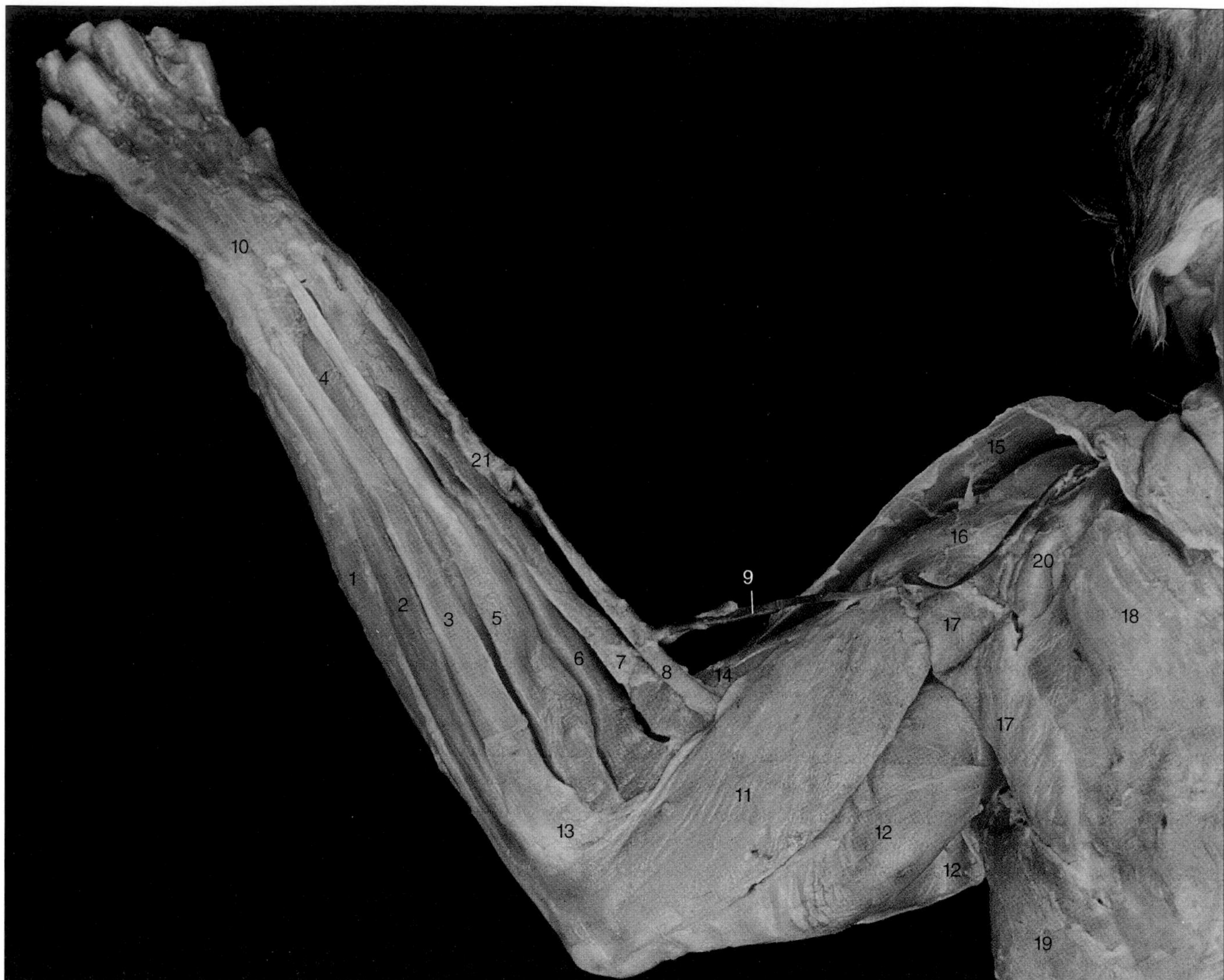

Figure 4-7
Superficial Muscles of the Cat Left Thoracic Limb, Dorsal View

1. Flexor carpi ulnaris muscle
2. Extensor carpi ulnaris muscle
3. Extensor carpi digitorum lateralis muscle
4. Extensor pollicis brevis muscle
5. Extensor digitorum communis muscle
6. Extensor carpi radialis brevis muscle
7. Extensor carpi radialis longus muscle
8. Brachioradialis muscle
9. Cephalic vein
10. Extensor retinaculum (dorsal carpal ligament)
11. Triceps brachii muscle (lateral head)
12. Triceps brachii muscle (long head)
13. Anconeus muscle
14. Brachialis muscle
15. Clavobrachialis muscle
16. Acromiodeltoid muscle
17. Spinodeltoid muscle
18. Acromiotrapezius muscle
19. Latissimus dorsi muscle
20. Levator scapulae ventralis
21. Antebrachial fascia

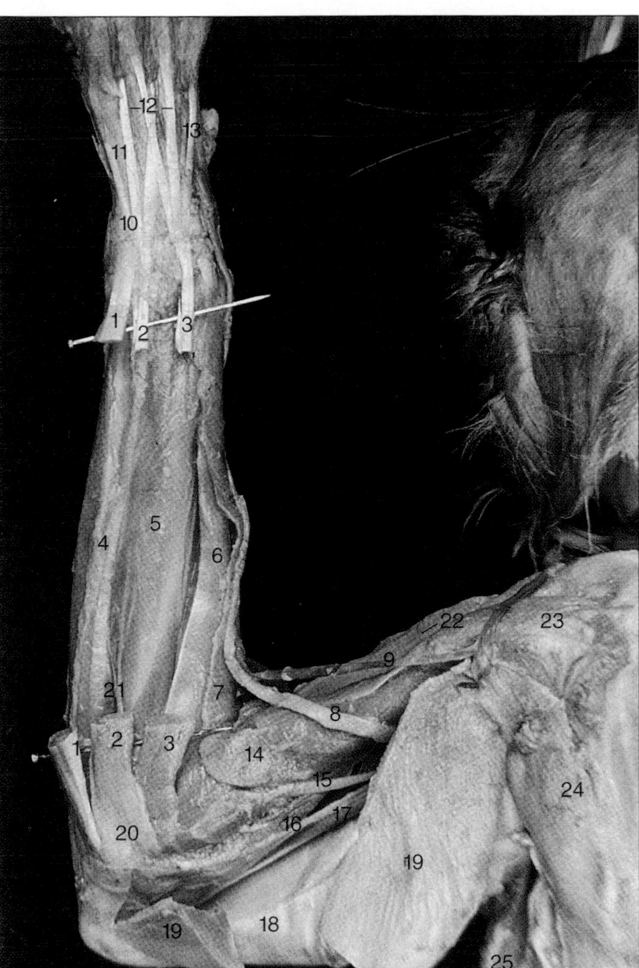

Figure 4-8

Deep Muscles of the Cat Left Thoracic Limb, Dorsal View

1. Extensor carpi ulnaris muscle (cut)
2. Extensor digitorum lateralis muscle (cut)
3. Extensor digitorum communis muscle (cut)
4. Extensor indicis proprius muscle
5. Extensor pollicis brevis muscle
6. Extensor carpi radialis muscle
7. Brachioradialis muscle
8. Radial nerve
9. Cephalic vein
10. Extensor retinaculum (dorsal carpal ligament)
11. Extensor digiti minimi tendon
12. Extensor digitorum tendons
13. Extensor indicis tendon
14. Brachioradialis muscle
15. Median nerve
16. Ulnar nerve
17. Triceps brachii muscle (medial head)
18. Triceps brachii muscle (long head)
19. Triceps brachii muscle (lateral head, cut)
20. Anconeus muscle
21. Posterior interosseous nerve
22. Clavobrachialis muscle
23. Acromiodeltoid muscle
24. Spinodeltoid muscle
25. Latissimus dorsi muscle

Figure 4-9
Superficial Muscles of the Cat Thorax, Ventral View

1. Clavobrachialis muscle
2. Pectoantebrachialis muscle
3. Pectoralis major muscle
4. Pectoralis minor muscle
5. Xiphihumeralis muscle
6. Epitrochlearis muscle
7. Latissimus dorsi muscle
8. External oblique muscle
9. Rectus abdominis muscle (deep to aponeurosis)
10. Linea alba
11. Inferior angle of scapula
12. Teres major muscle
13. Subscapularis muscle

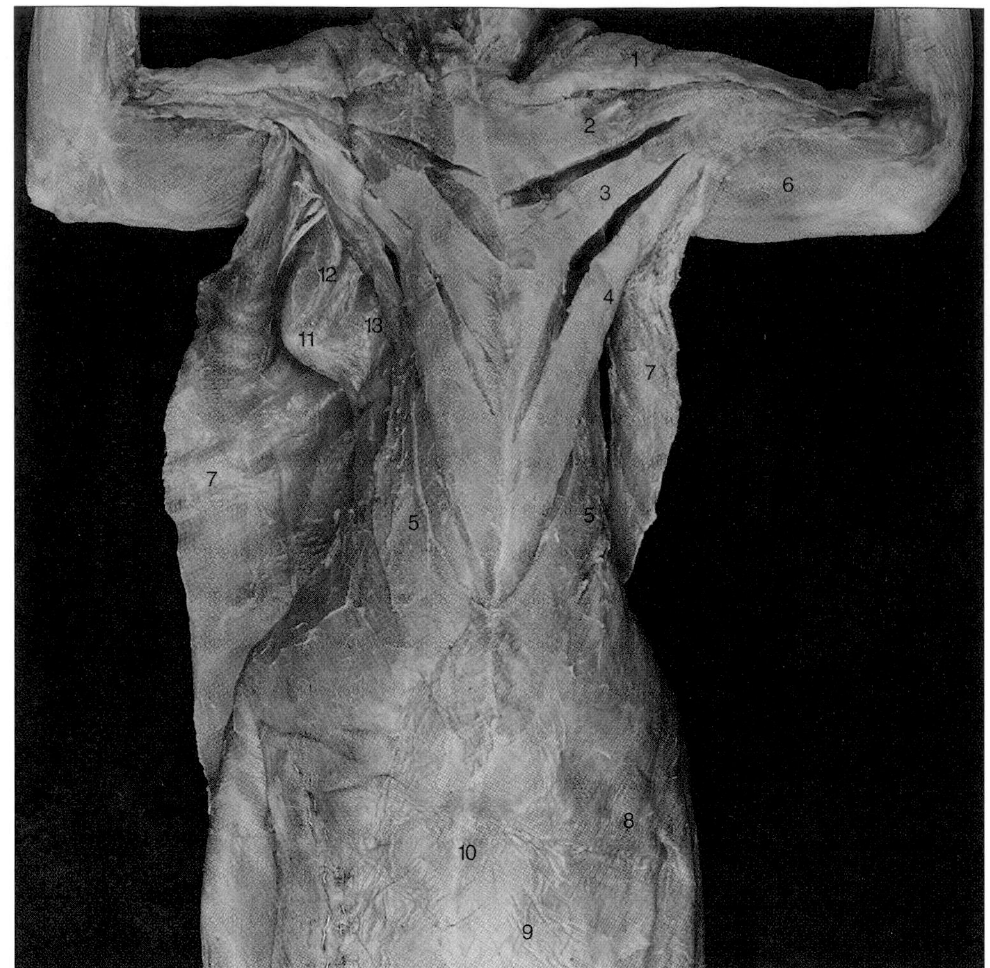

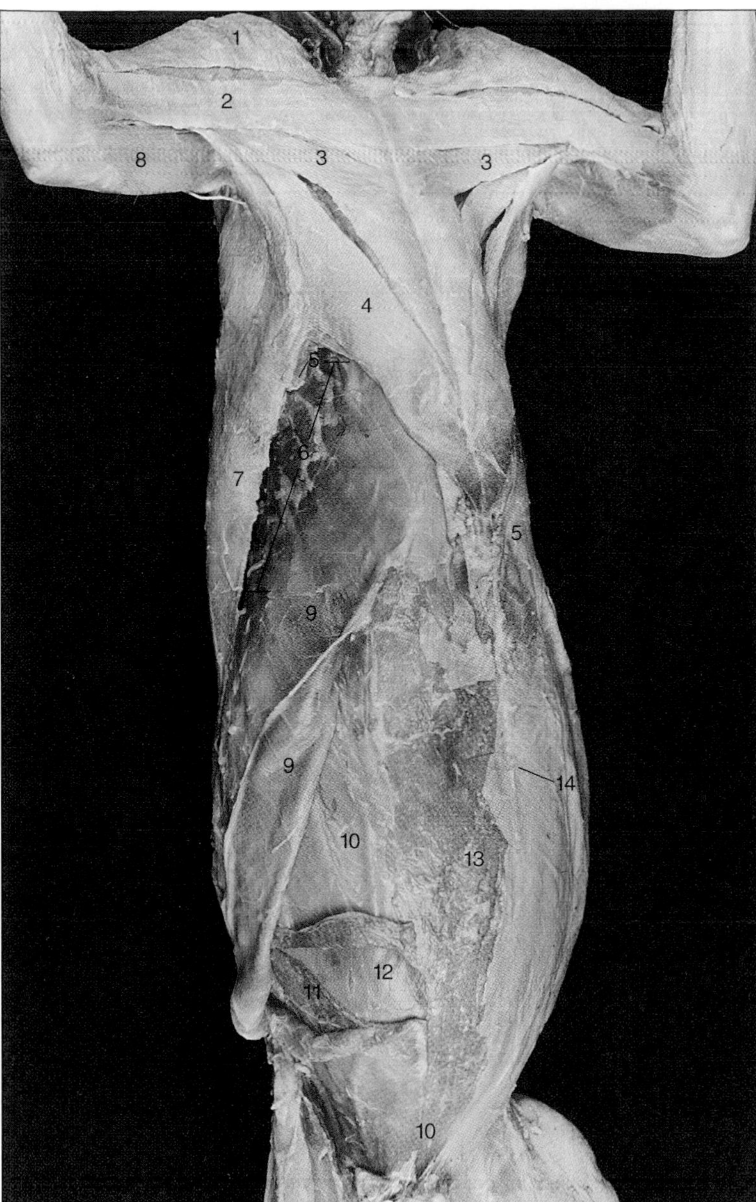

Figure 4-10
Superficial Muscles of the Cat, Abdomen and Thorax

1. Clavobrachialis muscle
2. Pectoantebrachialis muscle
3. Pectoralis major muscle
4. Pectoralis minor muscle
5. Xiphihumeralis muscle (removed on right)
6. Serratus anterior muscle
7. Latissimus dorsi muscle (cut to reveal underlying muscles)
8. Epitrochlearis muscle
9. External oblique muscle (partially reflected)
10. Internal oblique muscle (partially reflected)
11. Transversalis abdominis muscle
12. Peritoneum
13. Rectus abdominis muscle
14. Linea alba

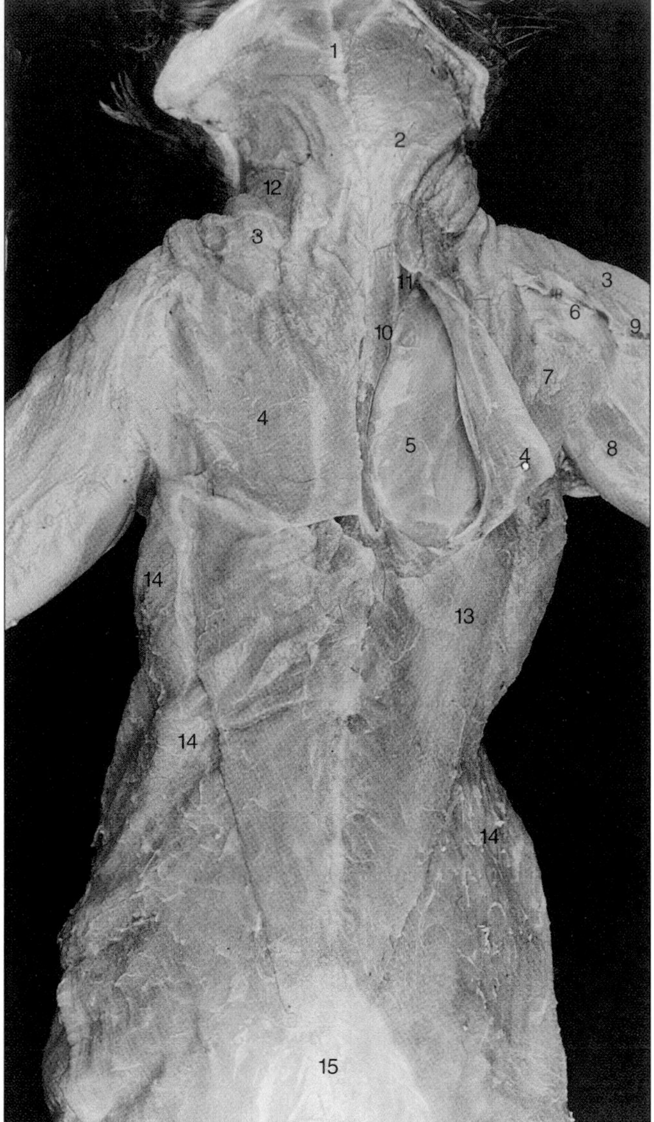

Figure 4-11
Superficial Muscles of the Cat Neck and Back

1. Nuchal ligament
2. Clavotrapezius muscle
3. Clavobrachialis muscle
4. Acromiotrapezius muscle (cut and reflected on right)
5. Supraspinatus muscle
6. Acromiodeltoid muscle
7. Spinodeltoid muscle
8. Triceps brachii muscle (long head)
9. Cephalic vein
10. Rhomboideus minor muscle
11. Rhomboideus capitis muscle (occipitoscapularis muscle)
12. Splenius capitis muscle
13. Spinotrapezius muscle
14. Latissimus dorsi muscle
15. Lumbodorsal fascia

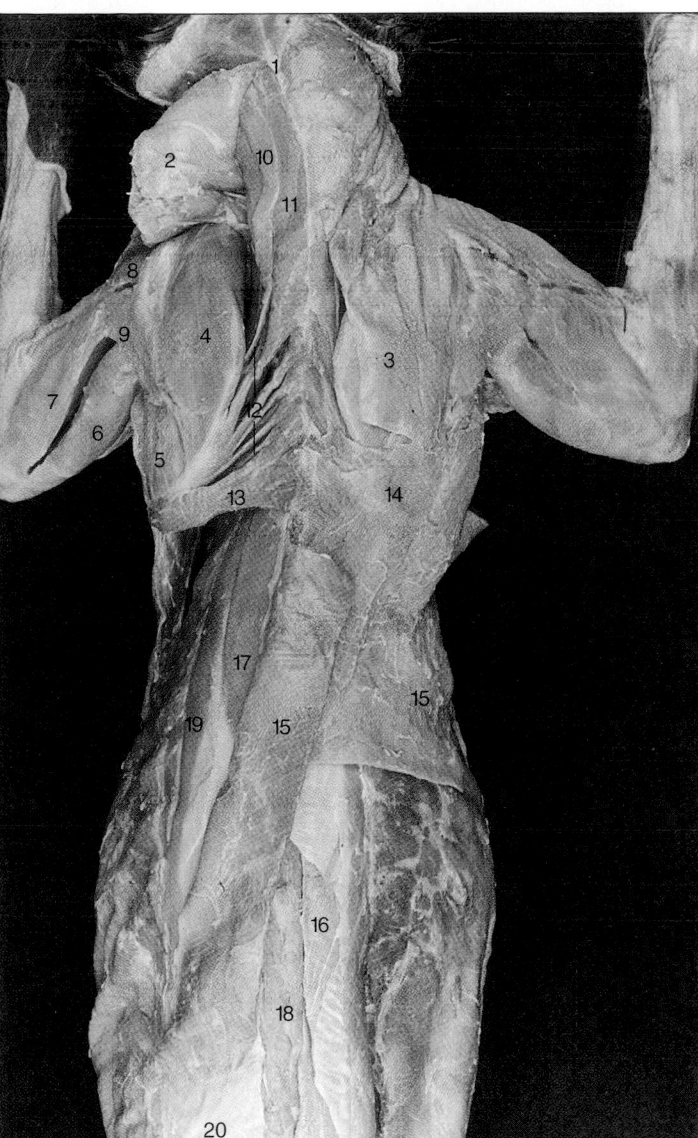

Figure 4-12
Deep Muscles of the Cat Neck and Back

1. Nuchal ligament
2. Clavotrapezius muscle (reflected on left)
3. Acromiotrapezius muscle (cut, removed altogether on left)
4. Supraspinatus muscle
5. Infraspinatus muscle
6. Triceps brachii muscle (long head)
7. Triceps brachii muscle (lateral head)
8. Acromiodeltoid muscle
9. Spinodeltoid muscle
10. Rhomboideus capitis muscle
11. Splenius capitis muscle
12. Rhomboideus minor muscle
13. Rhomboideus major muscle
14. Spinotrapezius muscle
15. Latissimus dorsi muscle (reflected on left, partially removed on right)
16. Multifidus muscle
17. Spinalis muscle
18. Longissimus muscle
19. Iliocostalis muscle
20. Lumbodorsal fascia (largely removed)

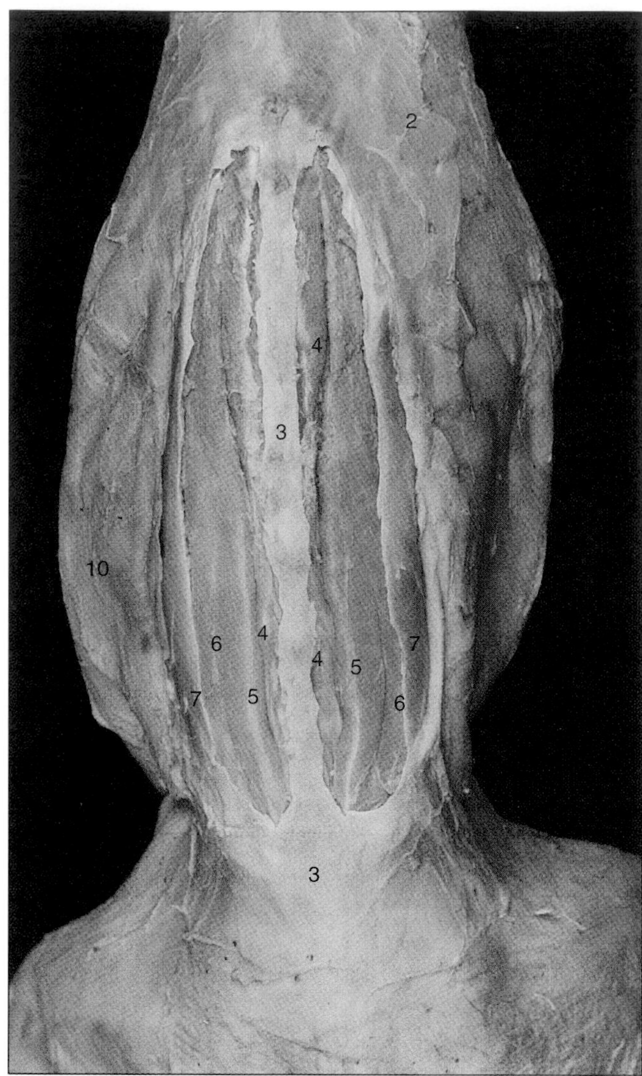

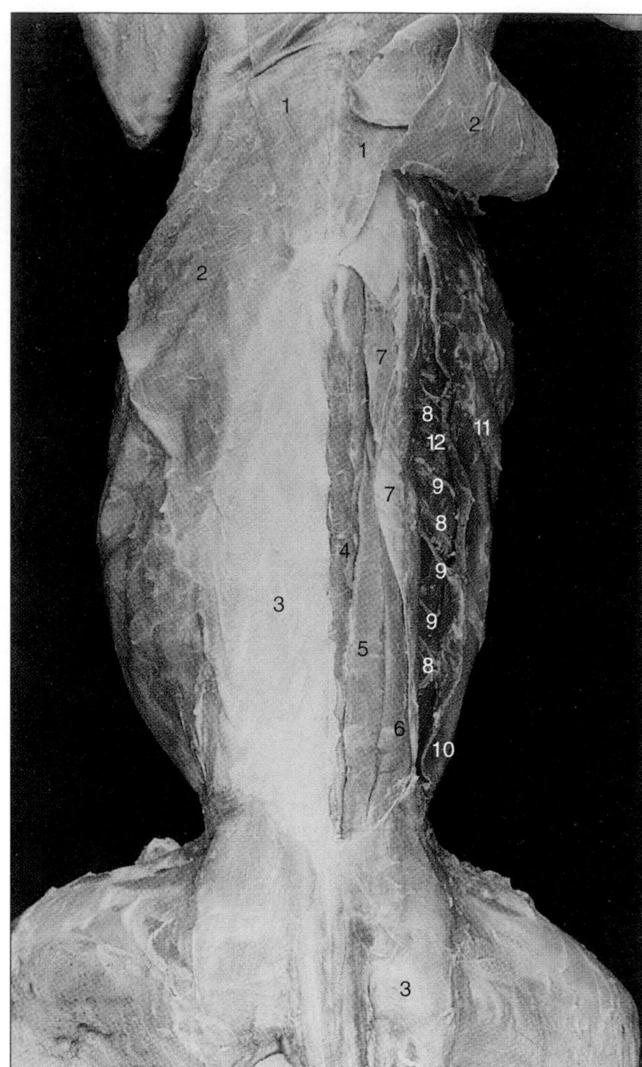

Figure 4-13
Deep Muscles of the Back of the Cat

1. Spinotrapezius muscle
2. Latissimus dorsi muscle (cut and rolled on right)
3. Lumbodorsal fascia
4. Multifidus muscle
5. Spinalis muscle
6. Longissimus muscle
7. Iliocostalis muscle
8. Rib
9. Dorsal ramus of spinal nerve
10. External oblique muscle
11. External intercostal muscle
12. Internal intercostal muscle

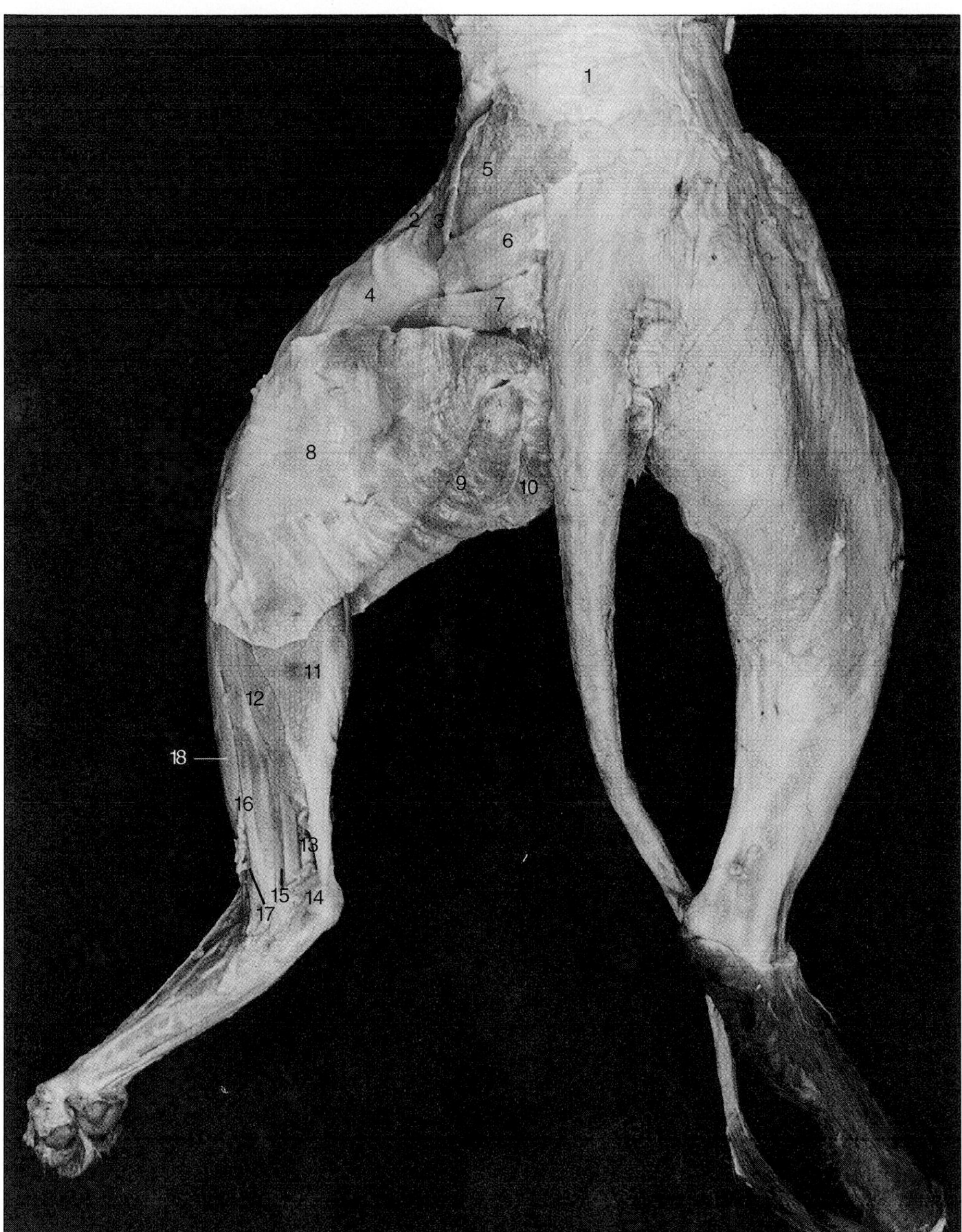

Figure 4-14
Superficial Muscles of the Cat Left Hind Limb, Dorsal View

1. Lumbodorsal fascia
2. Sartorius muscle
3. Tensor fascia latae muscle
4. Iliotibial tract
5. Gluteus medius muscle
6. Gluteus maximus muscle
7. Caudofemoralis muscle
8. Biceps femoris muscle
9. Semitendinosus muscle
10. Semimembranosus muscle
11. Gastrocnemius muscle
12. Soleus muscle
13. Achilles tendon
14. Calcaneal tuberosity
15. Flexor hallucis longus muscle
16. Peroneus brevis muscle
17. Peroneus longus muscle and tendon
18. Tibialis anterior muscle

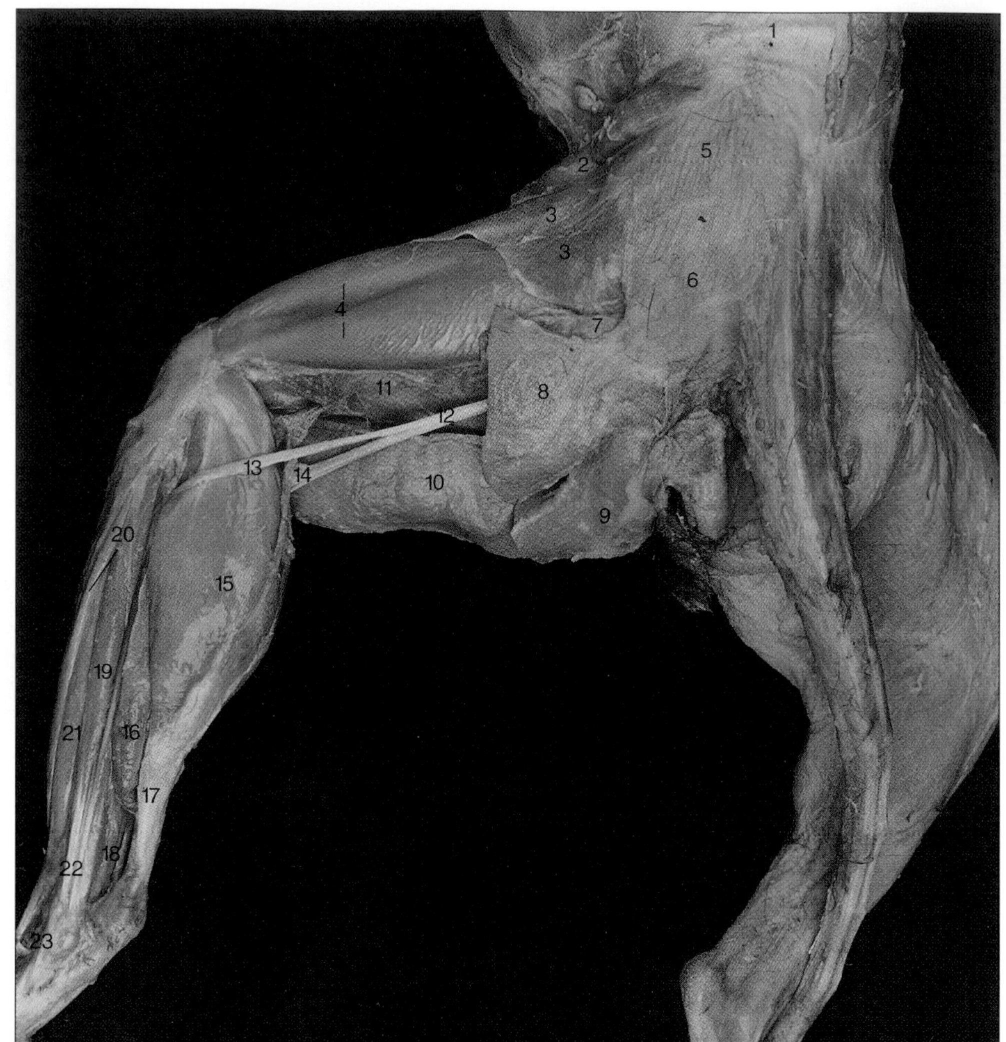

Figure 4-15
Deep Muscles of the Cat Left Hind Limb, Dorsal View

1. Lumbodorsal fascia
2. Sartorius muscle
3. Tensor fascia latae muscle
4. Vastus lateralis muscle
5. Gluteus medius muscle (under fascia)
6. Gluteus maximus muscle (under fascia)
7. Caudofemoralis muscle
8. Biceps femoris muscle (cut)
9. Semitendinosus muscle (cut)
10. Semimembranosus muscle
11. Adductor femoris muscle
12. Sciatic nerve
13. Common peroneal division of sciatic nerve
14. Tibial division of sciatic nerve
15. Gastrocnemius muscle
16. Soleus muscle
17. Achilles tendon
18. Flexor hallucis longus muscle
19. Peroneus longus muscle
20. Tibialis anterior muscle
21. Extensor digitorum longus muscle
22. Proximal extensor retinaculum
23. Distal extensor retinaculum

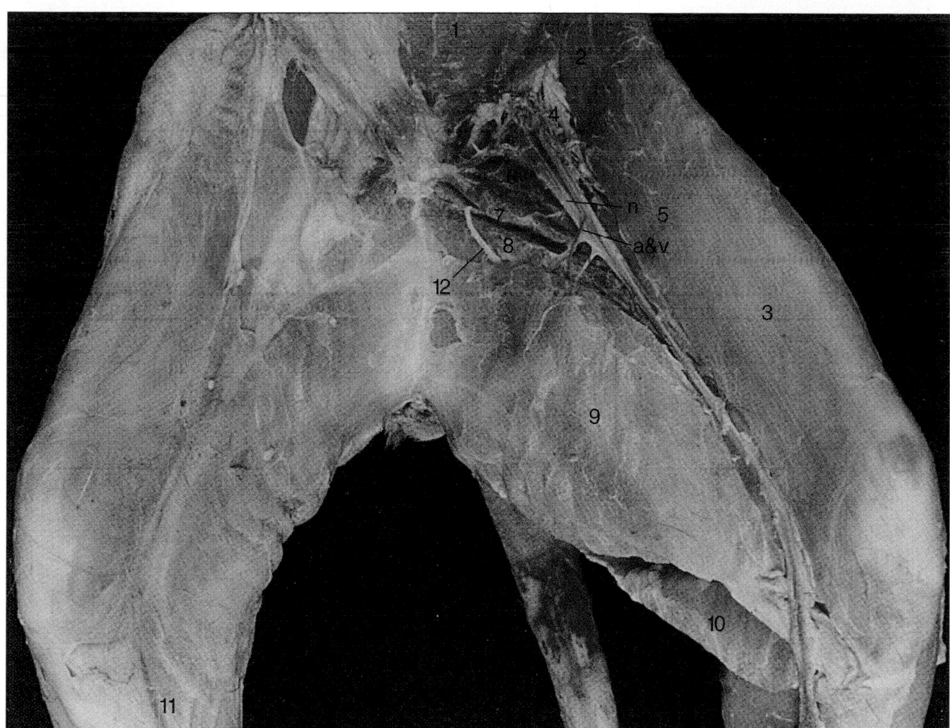

Figure 4-16
Superficial Muscles of the Cat Left Hind Limb, Medial View

1. Rectus abdominis muscle
2. External oblique muscle
3. Sartorius muscle
4. Iliopsoas muscle (deep to blood vessels)
5. Femoral artery (a), vein (v), and nerve (n)
6. Pectineus muscle (deep to blood vessels)
7. Adductor longus muscle
8. Adductor femoris muscle
9. Gracilis muscle
10. Semitendinosus muscle
11. Greater saphenous vein
12. Branch of obturator nerve

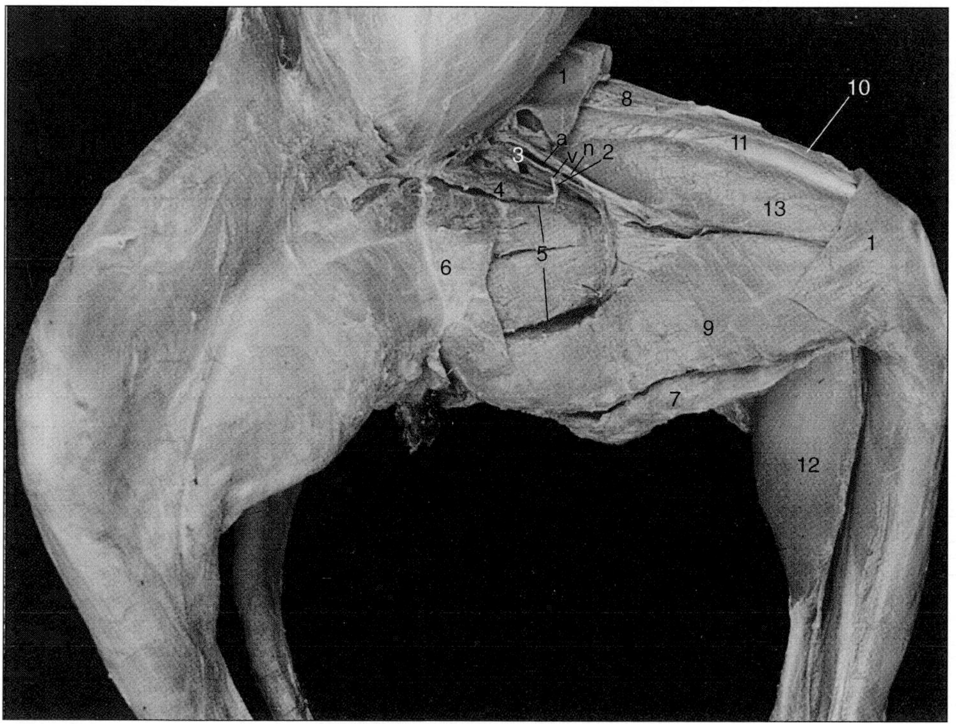

Figure 4-17
Superficial Muscles of the Cat Left Hind Limb, Medial View

1. Sartorius muscle (cut)
2. Femoral artery (a), vein (v), and nerve (n)
3. Pectineus muscle (deep to blood vessels)
4. Adductor longus muscle
5. Adductor femoris muscle
6. Gracilis muscle (cut)
7. Semitendinosus muscle
8. Tensor fascia latae muscle
9. Semimembranosus muscle
10. Vastus lateralis muscle
11. Rectus femoris muscle
12. Gastrocnemius muscle
13. Vastus medialis muscle

Figure 4-18
Deep Muscles of the Cat Left Hind Limb, Medial View

1. Sartorius muscle (cut)
2. Tensor fascia latae muscle
3. Vastus lateralis muscle
4. Rectus femoris muscle
5. Femoral artery (a), vein (v), and nerve (n)
6. Middle caudal femoral artery and vein
7. Pectineus muscle (under femoral artery and vein)
8. Adductor longus muscle
9. Adductor femoris muscle
10. Gracilis muscle (cut)
11. Semimembranosus muscle
12. Semitendinosus muscle
13. Biceps femoris muscle
14. Gastrocnemius muscle (reflected)
15. Soleus muscle
16. Achilles tendon
17. Posterior tibial nerve
18. Flexor hallucis longus muscle
19. Flexor digitorum longus muscle
20. Tibialis posterior muscle
21. Tibia
22. Tibialis anterior muscle
23. Proximal extensor retinaculum
24. Vastus medialis muscle

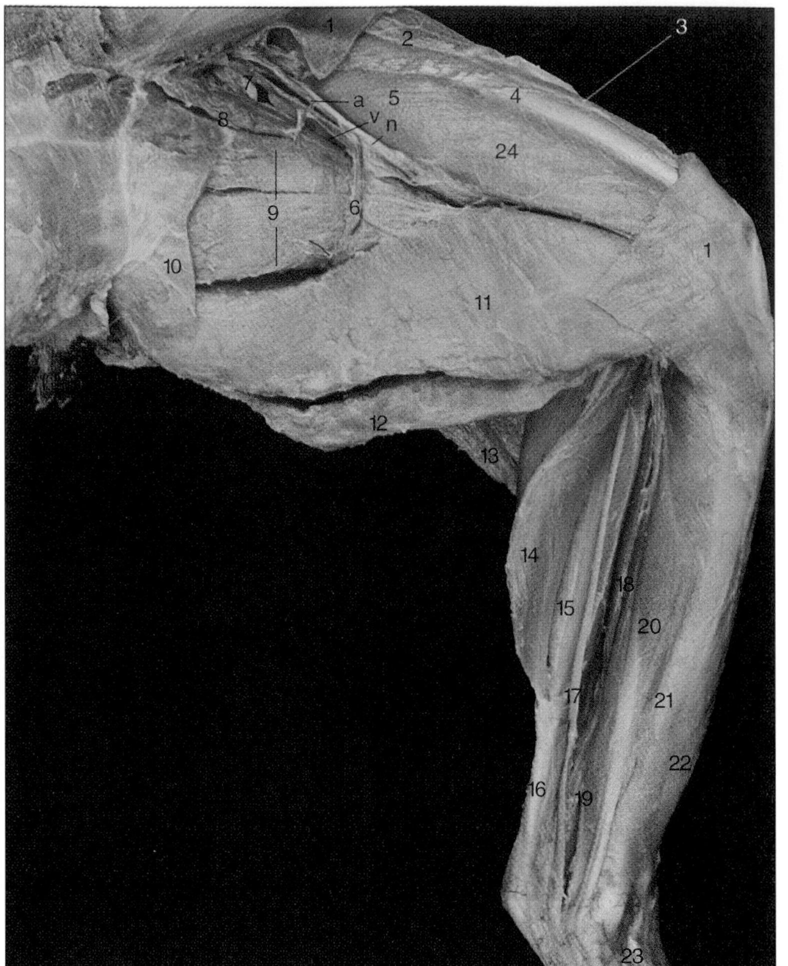

Figure 4-19
Deep Muscles of the Cat Shoulder and Thorax, Right Ventral View

1. Latissimus dorsi muscle (reflected)
2. Scalenus muscles
   a. Anterior (continuous with transversus costarum)
   b. Medius
   c. Posterior
3. Axillary artery (a) and vein (v)
4. Radial nerve
5. External jugular vein
6. Internal jugular vein
7. Thoracodorsal nerve
8. Long thoracic nerve
9. Thoracoacromial blood vessels
10. Serratus ventralis muscle
11. Teres major muscle
12. Subscapularis muscle
13. Sternum
14. Ventral thoracic nerve
15. Pectoral muscles

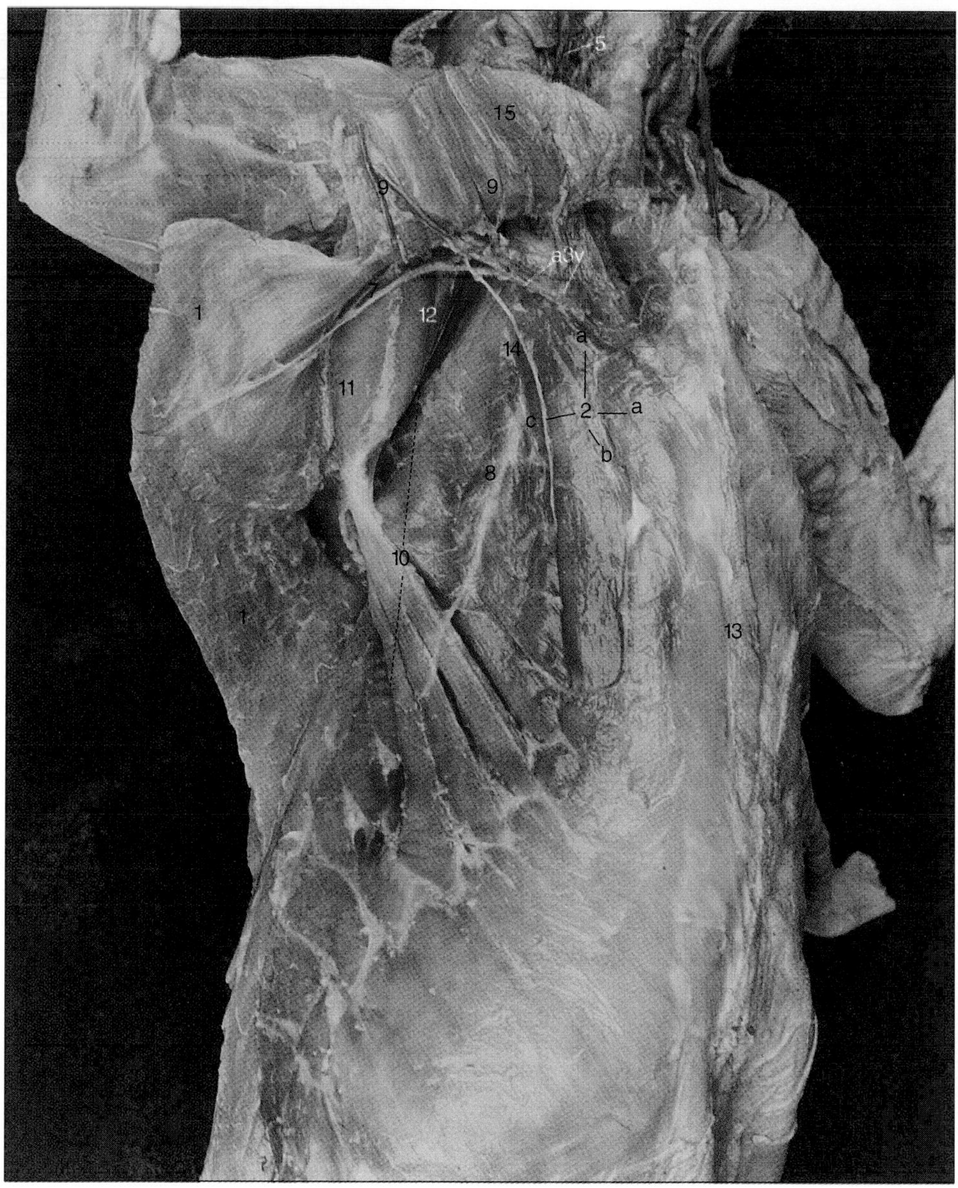

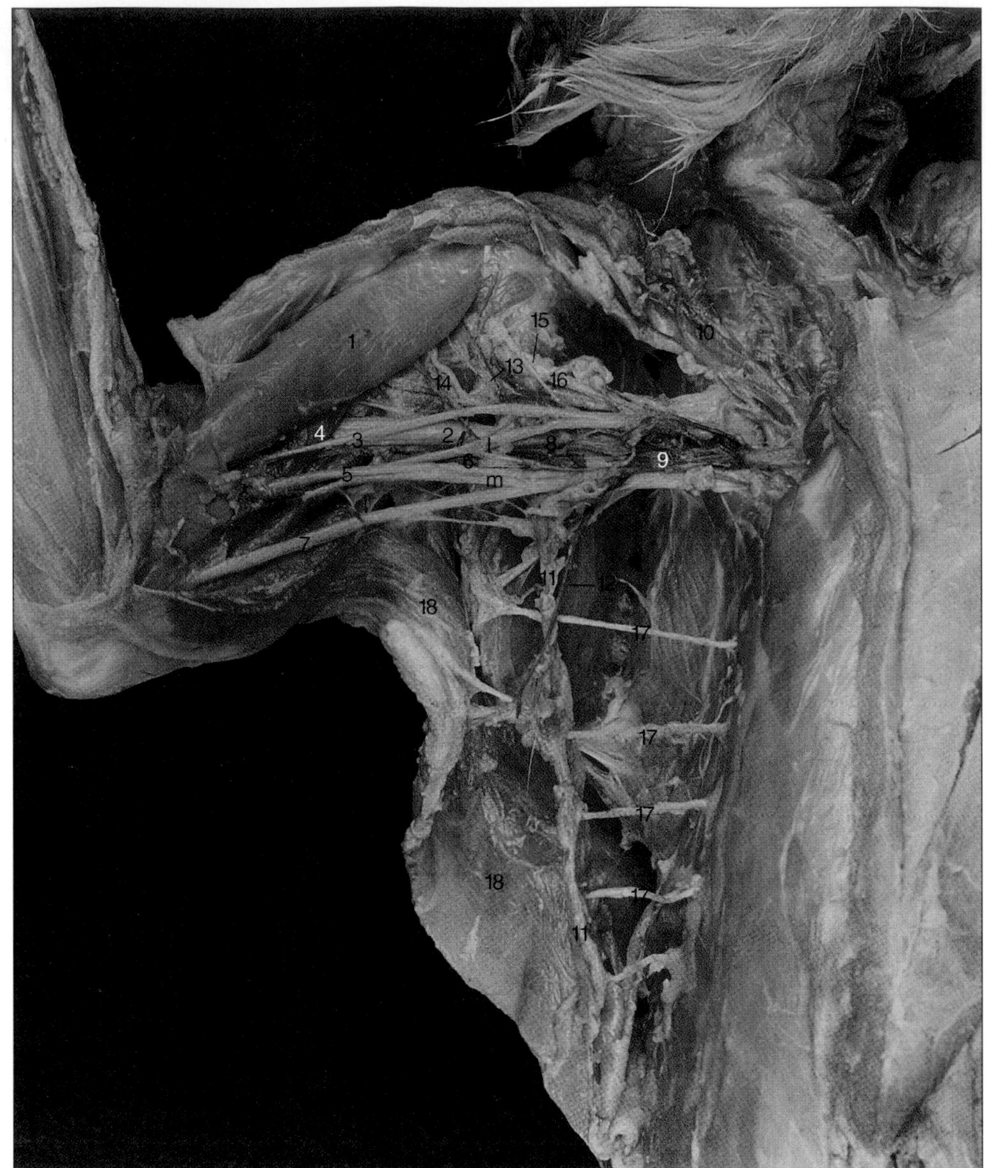

Figure 4-20
Brachial Plexus of the Cat, Right Ventral View

1. Biceps brachii muscle
2. Radial nerve
3. Musculocutaneous nerve
4. Coracobrachialis muscle
5. Median nerve
6. Lateral (l) and median (m) roots of the median nerve
7. Ulnar nerve
8. Axillary artery
9. Axillary vein
10. External jugular vein
11. Thoracodorsal nerve
12. Thoracodorsal artery
13. Thoracoacromial artery
14. Anterior circumflex humeral artery and axillary nerve
15. Caudal subscapular nerve
16. Proximal subscapular nerve
17. Dorsal rami of thoracic nerves
18. Latissimus dorsi muscle (reflected)

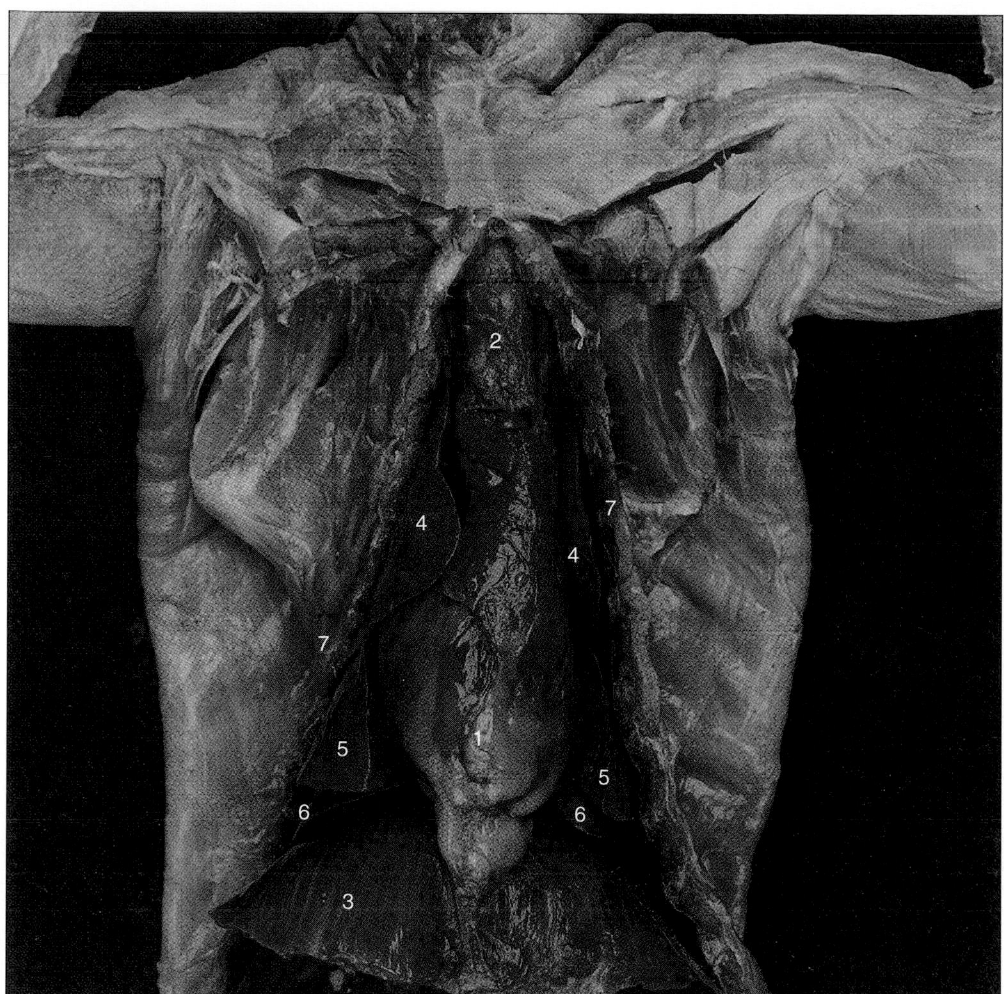

Figure 4-21
Thoracic Cavity of the Cat

1. Heart within pericardium
2. Thymus gland
3. Diaphragm
4. Lung, anterior lobe
5. Lung, middle lobe
6. Lung, posterior lobe
7. Ribs (cut)

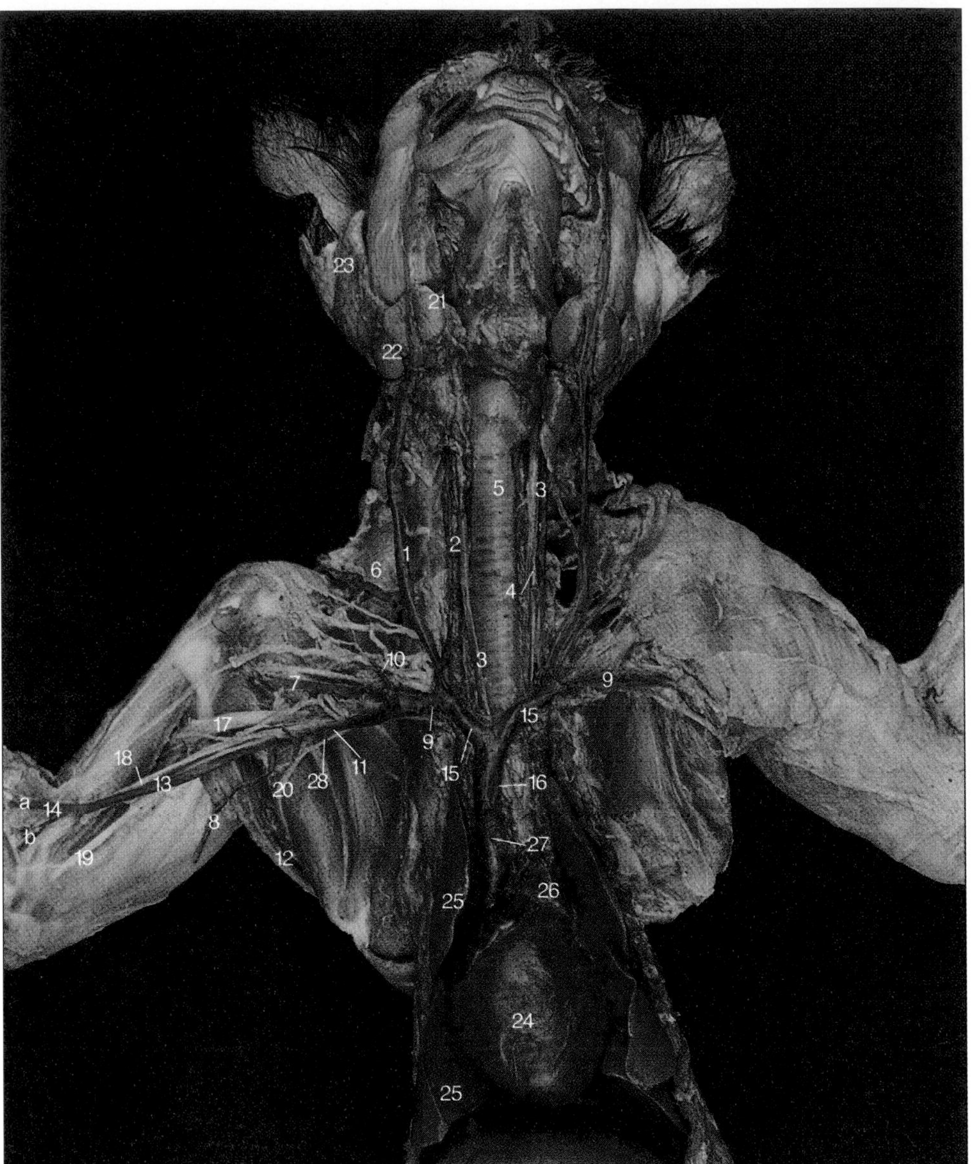

Figure 4-22
Major Veins of the Cat, Neck and Thorax

1. External jugular vein
2. Internal jugular vein
3. Common carotid artery (right)
4. Vagus nerve
5. Trachea
6. Transverse scapular vein
7. Subscapular vein
8. Thoracodorsal vein
9. Subclavian vein
10. Ventral thoracic (cut)
11. Axillary vein
12. Latissimus dorsi muscle
13. Brachial vein
14. a. radial vein
    b. ulnar vein
15. Brachiocephalic vein
16. Superior vena cava
17. Radial nerve
18. Median nerve
19. Ulnar nerve
20. Thoracodorsal nerve
21. Lymph node
22. Submandibular gland
23. Parotid gland
24. Heart
25. Lung
26. Thymus gland
27. Anterior thoracic vein (cut) (internal mammary vein)
28. Long thoracic vein (cut)

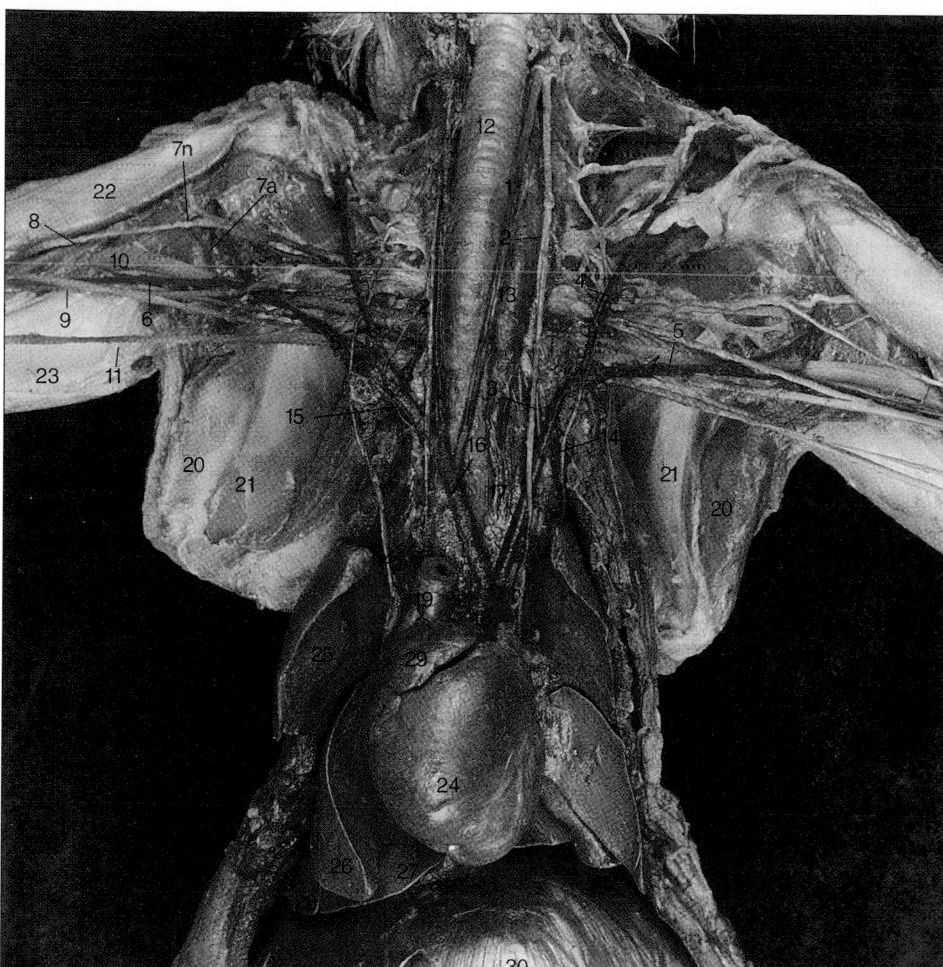

Figure 4-23
Major Arteries of the Cat, Neck and Thorax

1. Common carotid artery
2. Vagus nerve
3. Vertebral artery
4. Transverse scapular artery
5. Axillary artery
6. Brachial artery
7. Thoraco-acromial artery (a) and nerve (n)
8. Musculocutaneous nerve
9. Median nerve
10. Radial nerve
11. Ulnar nerve
12. Trachea
13. Esophagus (displaced to animal's left from normal position posterior to trachea)
14. Phrenic nerve
15. Right subclavian artery
16. Brachiocephalic artery
17. Left subclavian artery
18. Aortic arch
19. Anterior vena cava (cut)
20. Teres major muscle
21. Subscapularis muscle
22. Biceps brachii muscle
23. Triceps brachii muscle (long head)
24. Heart
25. Lung, anterior lobe
26. Lung, middle lobe
27. Lung, mediastinal lobe
28. Lung, posterior lobe
29. Right auricle
30. Diaphragm

Figure 4-24
Thorax of the Cat, Heart and Lungs Removed

1. Trachea
2. Esophagus
3. Aortic arch
4. Brachiocephalic artery
5. Right common carotid artery
6. Left common carotid artery
7. Vagus nerve
8. Sympathetic trunk
9. Left subclavian artery
10. Phrenic nerve
11. Vertebral artery
12. Left axillary artery
13. Thyrocervical artery
14. Internal mammary artery
15. Right and left primary bronchi
16. Inferior vena cava
17. Diaphragm
18. Rib
19. Subscapular artery (cut)

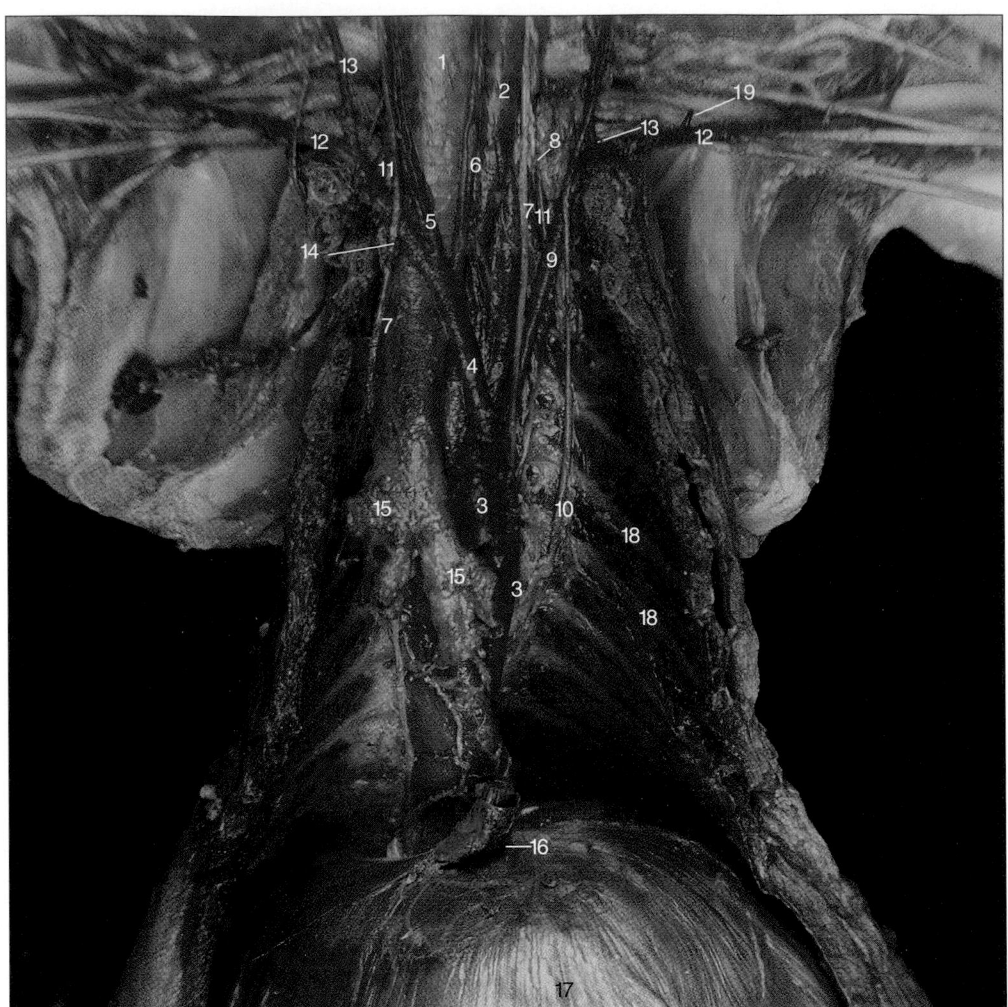

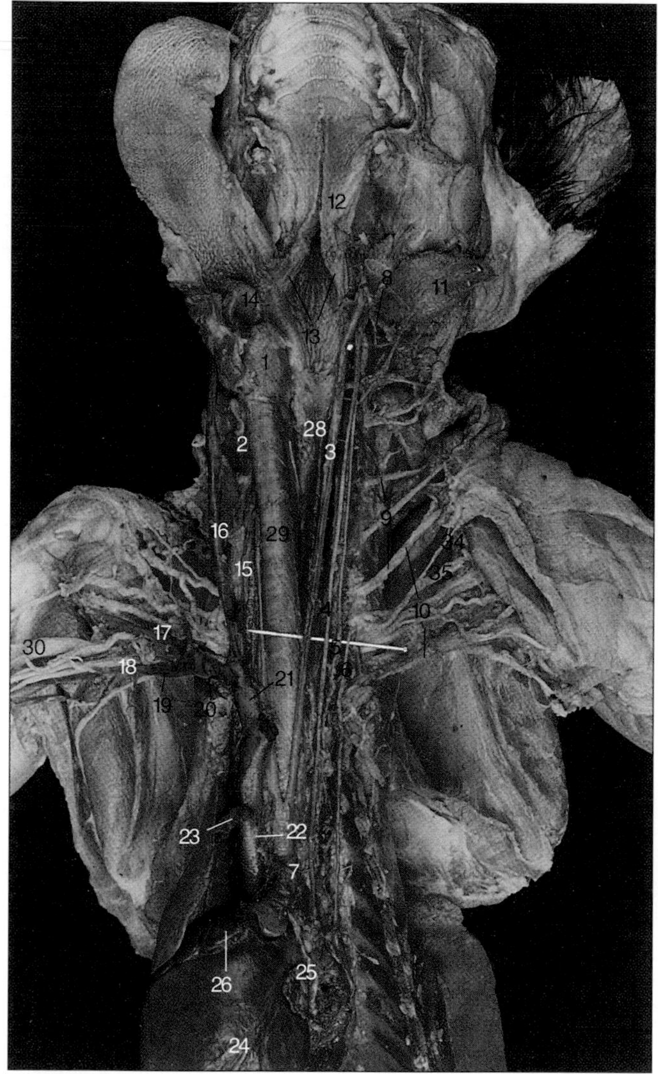

A

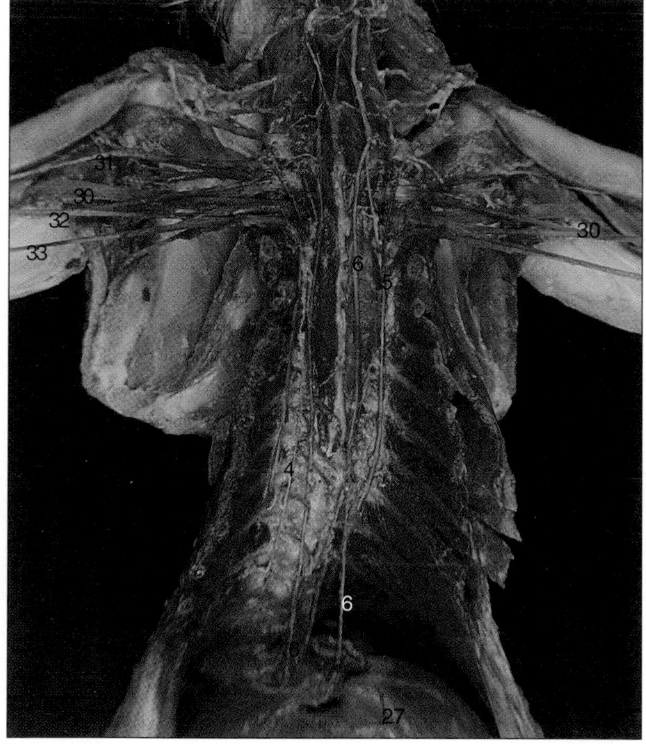

B

Figure 4-25
Veins, Arteries, and Nerves of the Cat Neck and Thorax
(A) Veins Removed on Cat's Left Side, Heart Reflected to Right
(B) Arteries, Veins, and Thoracic Viscera Removed

1. Larynx
2. Thyroid gland (reflected)
3. Common carotid artery
4. Vagus nerve
5. Sympathetic trunk (In **A,** two pins have been placed along sympathetic trunk. Upper pin head is just caudal to swelling of sympathetic trunk, the superior cervical ganglion. Lower transverse pin is just proximal to similar swelling, the middle cervical ganglion.)
6. Phrenic nerve
7. Aorta
8. Spinal accessory nerve (XI)
9. Spinal nerves IV, V, and VI
10. Brachial plexus
11. Lymph node (reflected and pinned)
12. Soft palate (cut)
13. Eustachian tubes (hidden behind reflected tissue of soft palate)
14. Epiglottis
15. Internal jugular vein
16. External jugular vein
17. Subscapular vein
18. Brachial vein
19. Axillary vein
20. Subclavian vein
21. Brachiocephalic vein
22. Superior vena cava
23. Azygous vein (cut)
24. Heart (reflected to cat's right)
25. Right auricle
26. Left auricle
27. Diaphragm
28. Esophagus
29. Trachea
30. Radial nerve
31. Musculocutaneous nerve
32. Median nerve
33. Ulnar nerve
34. Caudal subscapular nerve
35. Axillary nerve

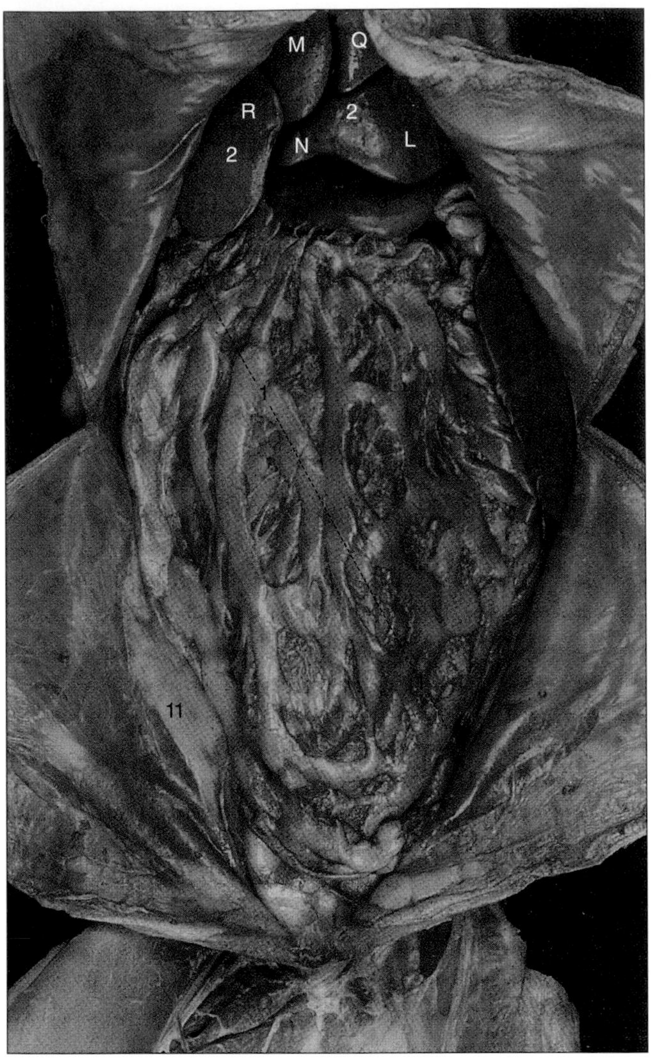

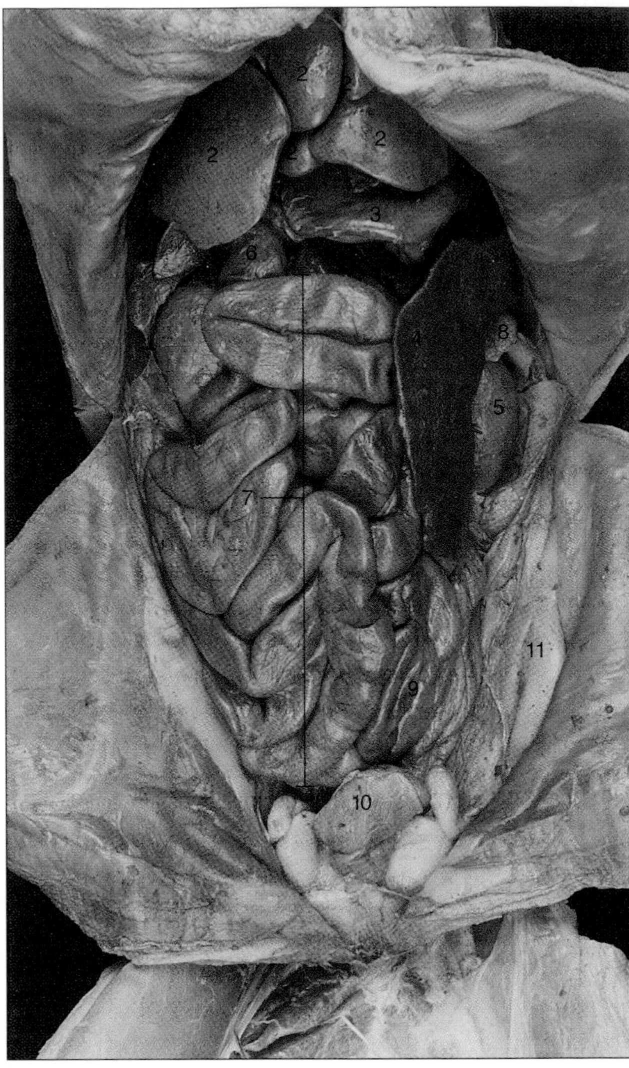

A

B

Figure 4-26
Abdominal Viscera of Cat with (A) Greater Omentum Intact and (B) Greater Omentum Removed

1. Greater omentum
2. Lobes of the liver
   R. Right lateral lobe
   M. Right medial lobe
   Q. Quadrate lobe
   N. Left medial lobe
   L. Left lateral lobe
3. Stomach (greater curvature)
4. Spleen
5. Kidney
6. Small intestine (duodenum)
7. Small intestine (jejuneum and ileum)
8. Pancreas
9. Large intestine (descending colon)
10. Urinary bladder
11. Abdominal fat

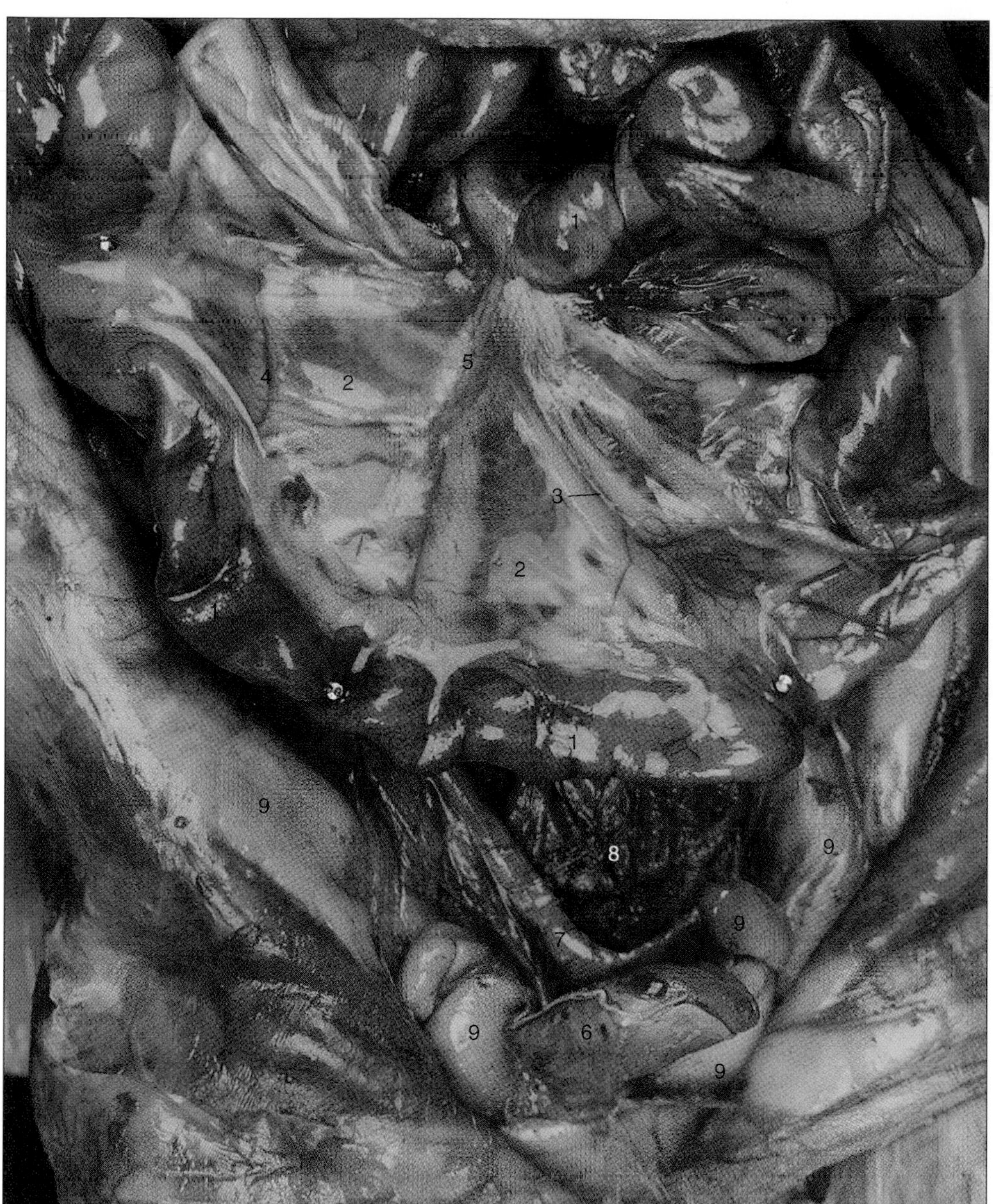

Figure 4-27
Abdominal Viscera of the Cat, Mesentery

1. Small intestine
2. Mesentery
3. Mesenteric artery
4. Mesenteric vein
5. Lymph vessel
6. Urinary bladder
7. Uterine horn
8. Descending colon
9. Abdominal fat

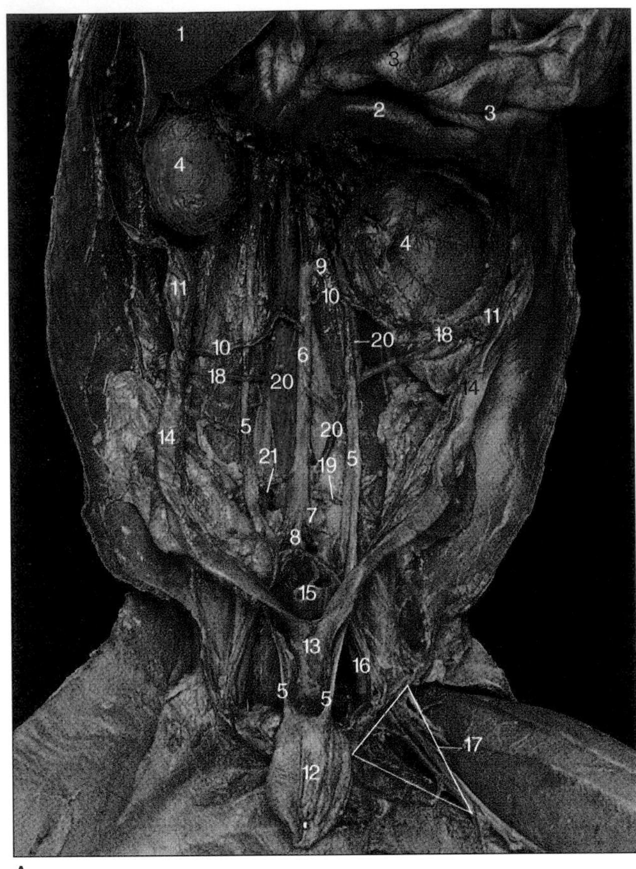

A

B

Figure 4-28
Urogenital System of the Female Cat (A) Nonpregnant and (B) Pregnant

1. Liver
2. Stomach (pylorus)
3. Small intestine
4. Kidney
5. Ureter
6. Abdominal aorta
7. External iliac artery
8. Common branch of internal iliac artery
9. Renal artery
10. Ovarian artery
11. Ovary
12. Urinary bladder (reflected and pinned)
13. Uterus
14. Uterine horn (in **B,** left horn contains two fetuses; right horn, three fetuses)
15. Rectum (cut in **A**)
16. External iliac artery and vein
17. Femoral triangle (containing femoral nerve, artery, and vein)
18. Ovarian vein
19. Iliolumbar artery and vein
20. Abdominal vena cava (split into two parallel vessels in **A**)
21. Iliolumbar artery
22. Pancreas
23. Spleen
24. Adrenal gland
25. Hepatic portal vein (cut)
26. Fetus
27. Placenta
28. Left gastroepiploic vein
29. Right gastroepiploic vein

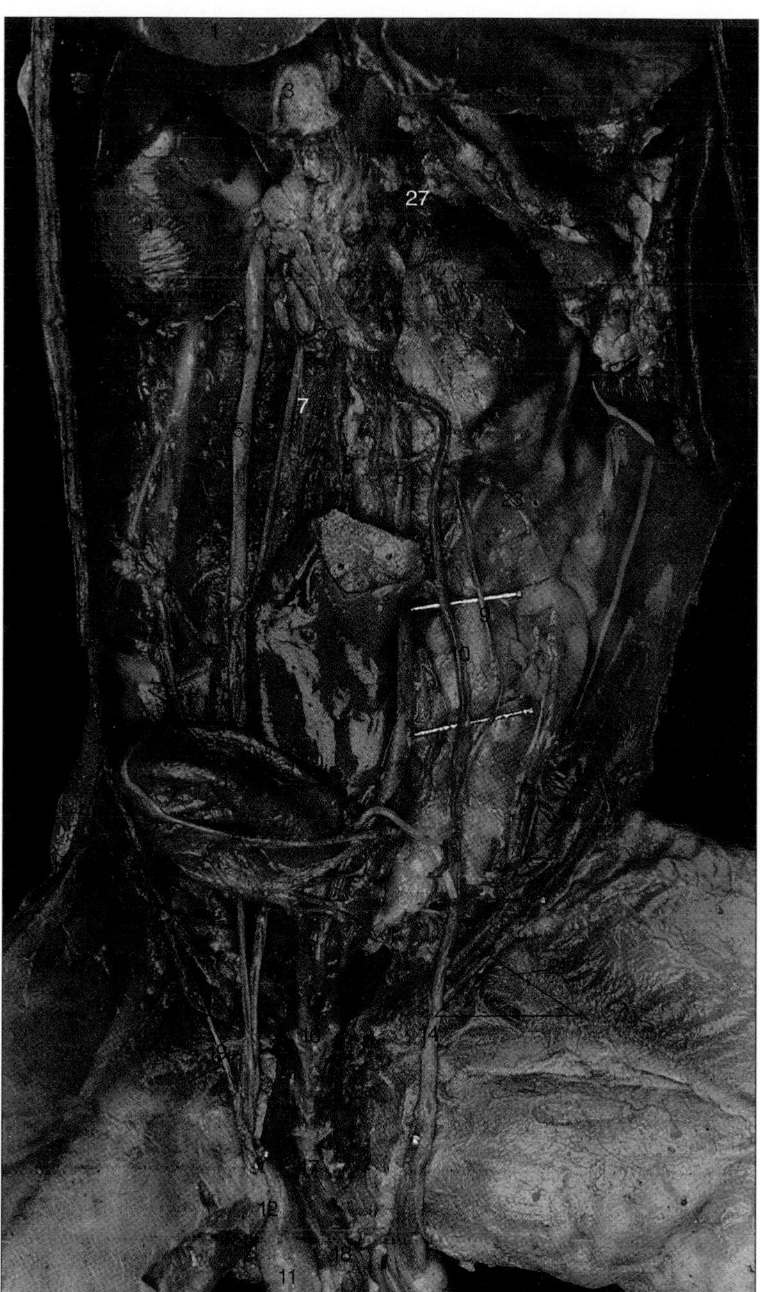

Figure 4-29
Urogenital System of the Male Cat

1. Liver
2. Stomach
3. Small intestine (duodenum, cut)
4. Kidney
5. Ureter
6. Abdominal aorta
7. Abdominal vena cava
8. Renal artery
9. Internal spermatic artery
10. Spermatic vein
11. Testis
12. Epididymis
13. Urinary bladder (reflected)
14. Vas deferens in spermatic cord
15. Urethra
16. Prostate gland
17. Bulbourethral (Cowper's) gland
18. Penis
19. Ligament of cremaster muscle
20. External inguinal ring
21. Femoral triangle
22. Rectum (cut)
23. Lumbar nerve (medial branch)
24. Umbilical (allantoic) artery
25. Spleen
26. Pancreas
27. Adrenal gland

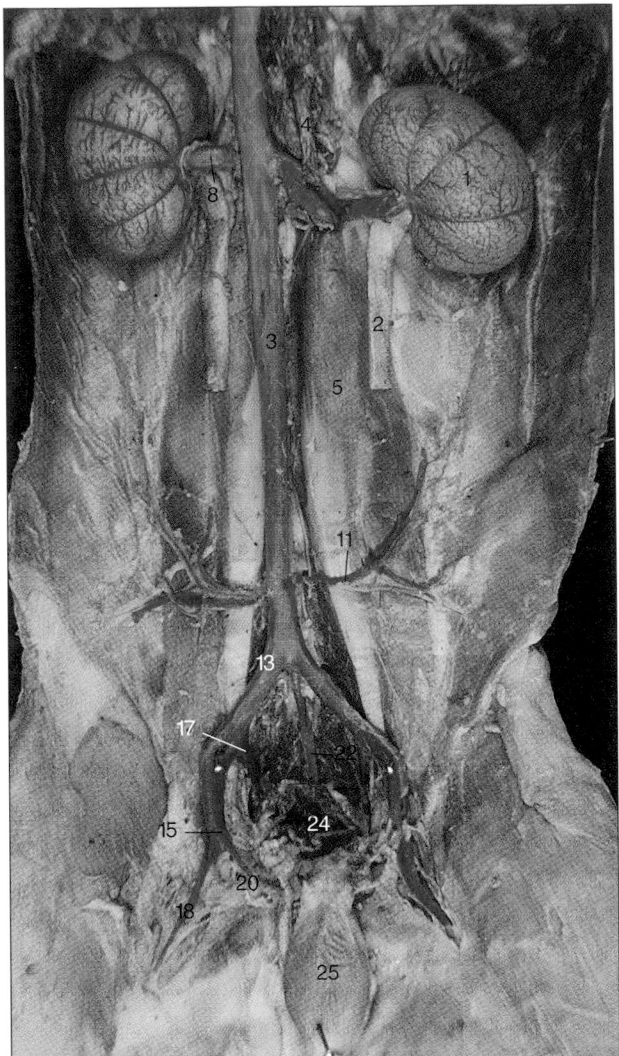

A

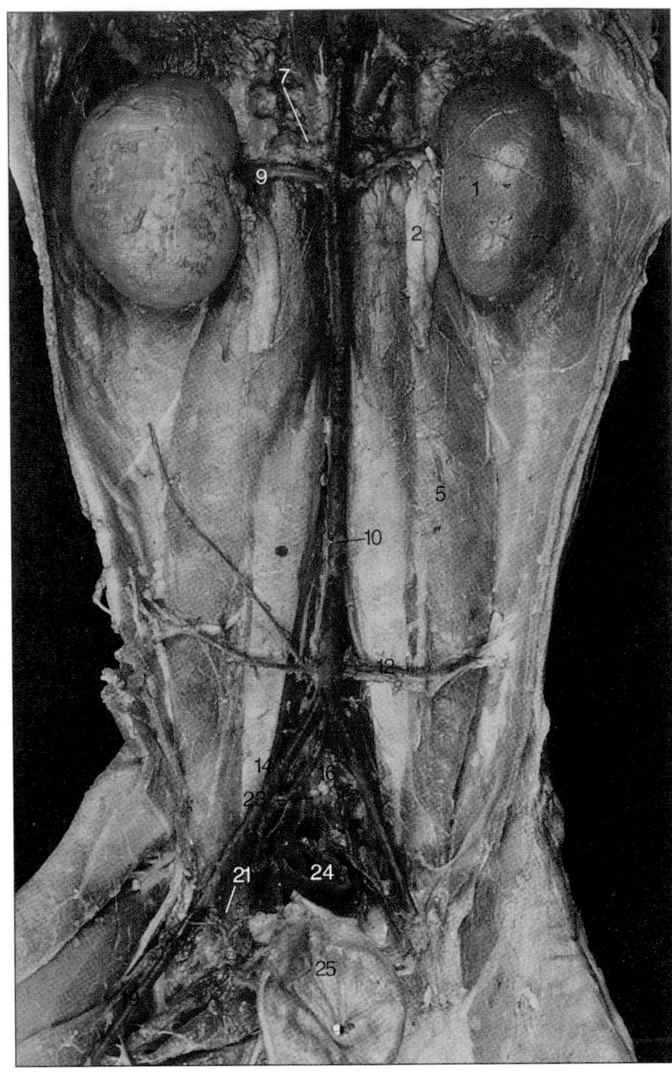

B

Figure 4-30

Major (A) Veins and (B) Arteries of the Cat Abdominopelvic Wall

1. Kidney
2. Ureter (cut and largely removed)
3. Abdominal vena cava (removed in **B**)
4. Abdominal aorta (cut and removed in **A**)
5. Psoas major and minor muscles
6. Celiac artery (cut and removed)
7. Suprarenal artery
8. Renal vein
9. Renal artery
10. Inferior mesenteric artery (cut and removed)
11. Iliolumbar vein
12. Iliolumbar artery
13. Common iliac vein
14. External iliac artery (no common iliac artery in cats)
15. External iliac vein
16. Internal iliac (hypogastric) artery
17. Internal iliac (hypogastric) vein
18. Femoral vein
19. Femoral artery
20. Deep femoral vein
21. Deep femoral artery
22. Caudal vein
23. Median sacral (caudal) artery
24. Rectum (cut)
25. Urinary bladder (reflected and pinned)

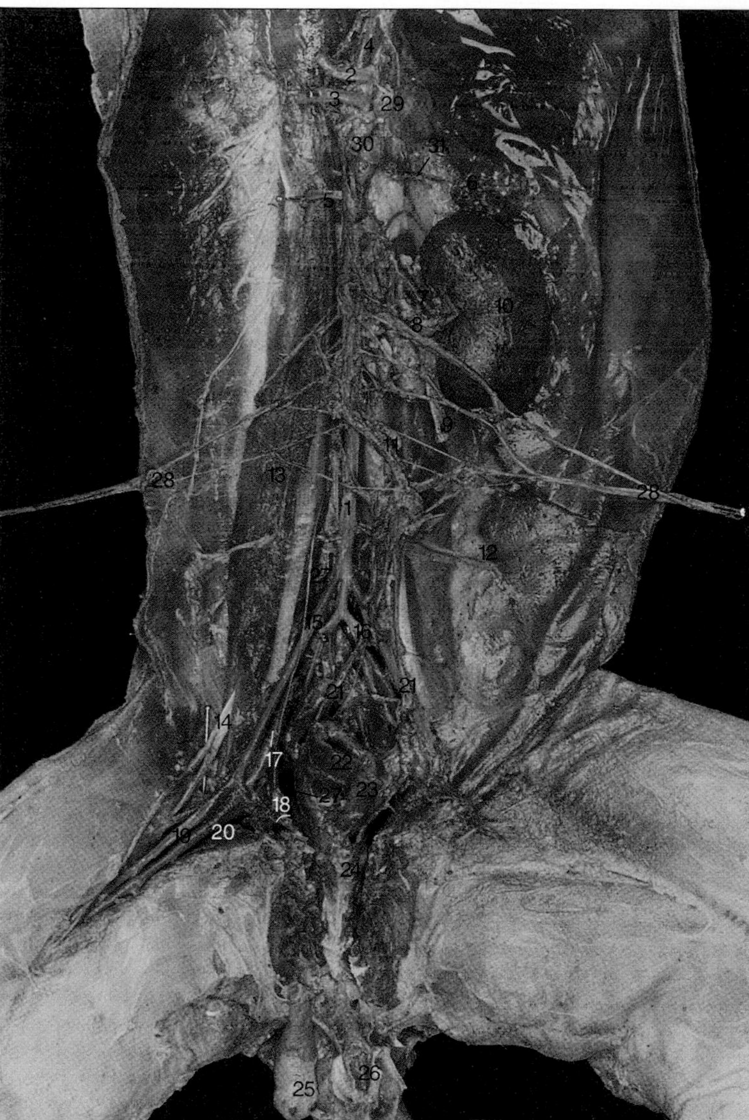

Figure 4-31
Nerves and Vessels of the Posterior Abdominopelvic Wall of the Cat

1. Abdominal aorta
2. Celiac artery (pinned)
3. Superior mesenteric artery (pinned)
4. Crus of diaphragm
5. Right suprarenal artery (cut)
6. Adrenal gland
7. Renal vein
8. Renal artery
9. Ureter (cut)
10. Kidney
11. Inferior mesenteric artery
12. Iliolumbar artery
13. Psoas major and minor muscles
14. Femoral nerve
15. External iliac artery
16. Internal iliac artery
17. External iliac vein
18. Deep femoral artery and vein
19. Femoral artery
20. Femoral vein
21. Spermatic artery
22. Rectum (cut)
23. Urethra (cut and urinary bladder removed)
24. Prostate gland
25. Testis
26. Penis
27. Genitofemoral nerve
28. Distribution of sympathetic trunk (pinned out bilaterally)
29. Celiac ganglion
30. Superior mesenteric ganglion
31. Left suprarenal artery and vein

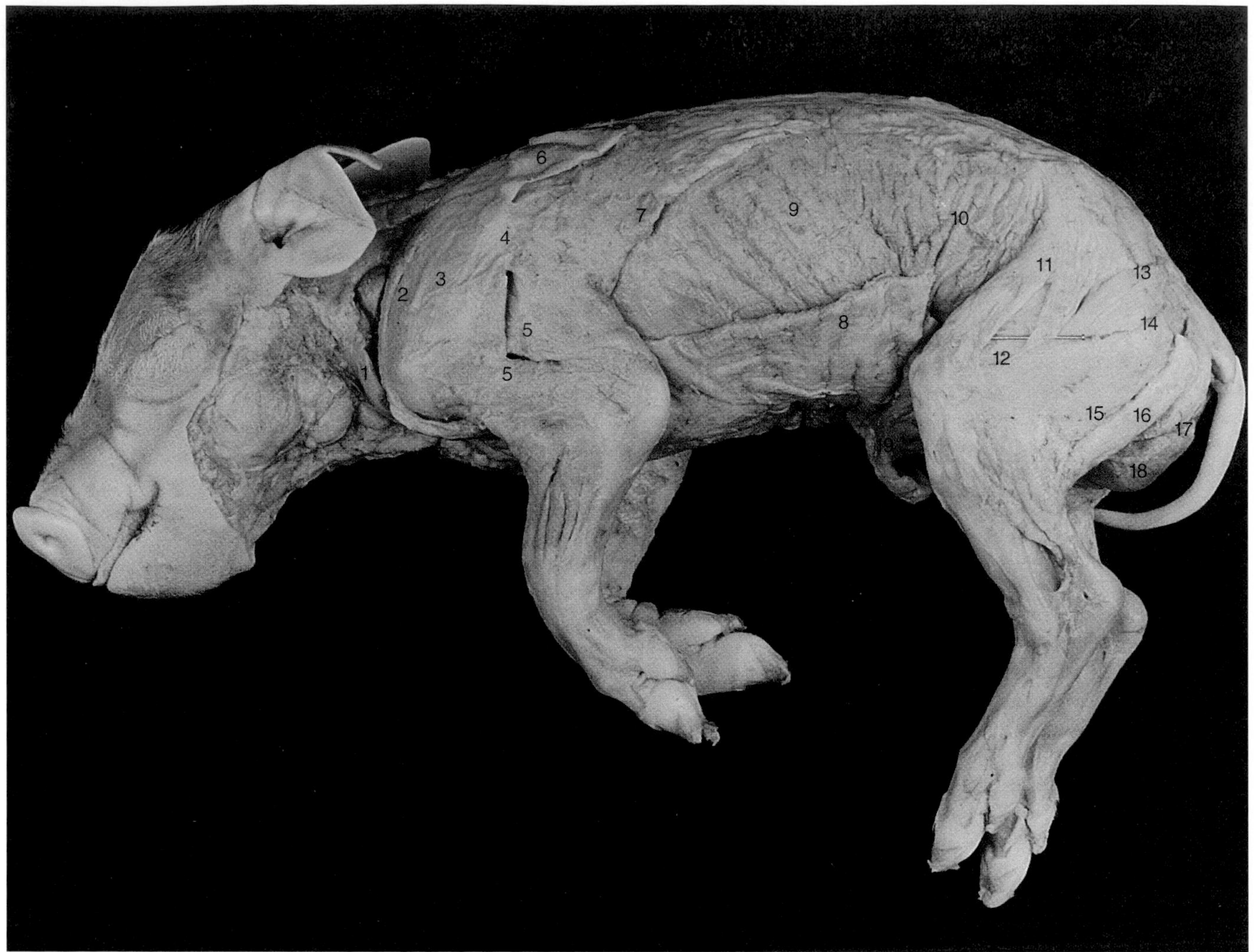

Figure 4-32
Superficial Muscles of the Fetal Pig, Left Lateral View

1. Clavotrapezius muscle
2. Clavobrachialis muscle
3. Acromiodeltoid muscle
4. Spinodeltoid muscle
5. Triceps brachii muscle
6. Spinotrapezius muscle (cut)
7. Latissimus dorsi muscle
8. External oblique muscle (cut)
9. Serratus anterior muscle
10. Internal oblique muscle
11. Tensor fascia latae muscle (split)
12. Vastus lateral muscle (under pin)
13. Gluteus medius muscle
14. Gluteus maximus muscle
15. Biceps femoris muscle
16. Semitendinosus muscle
17. Semimembranosus muscle
18. Testis
19. Umbilical cord

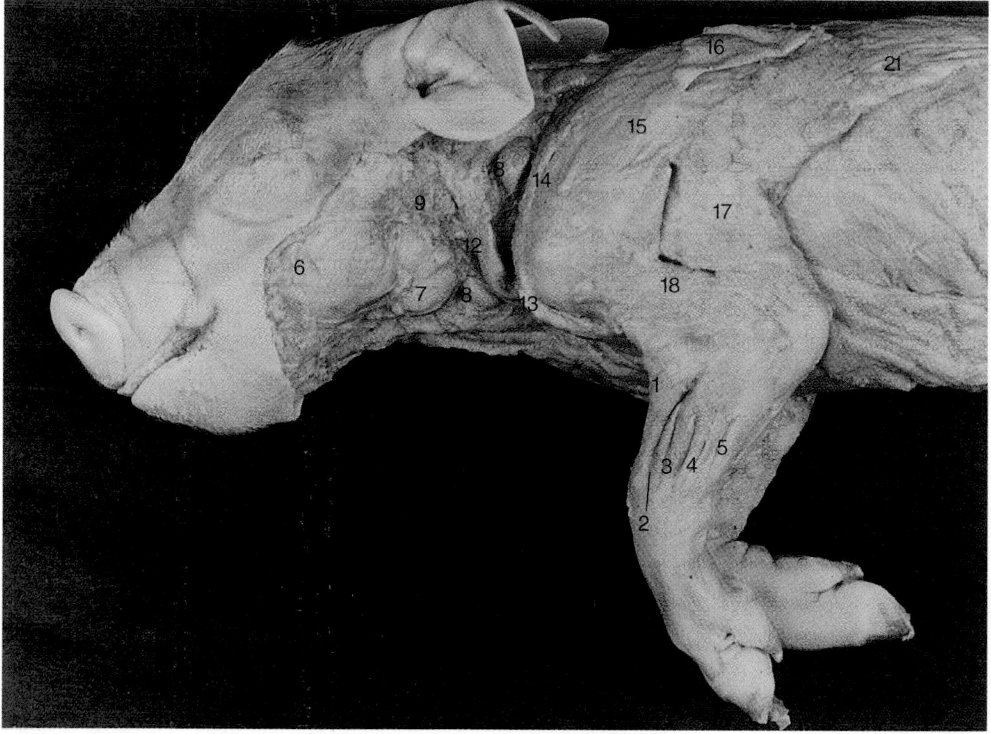

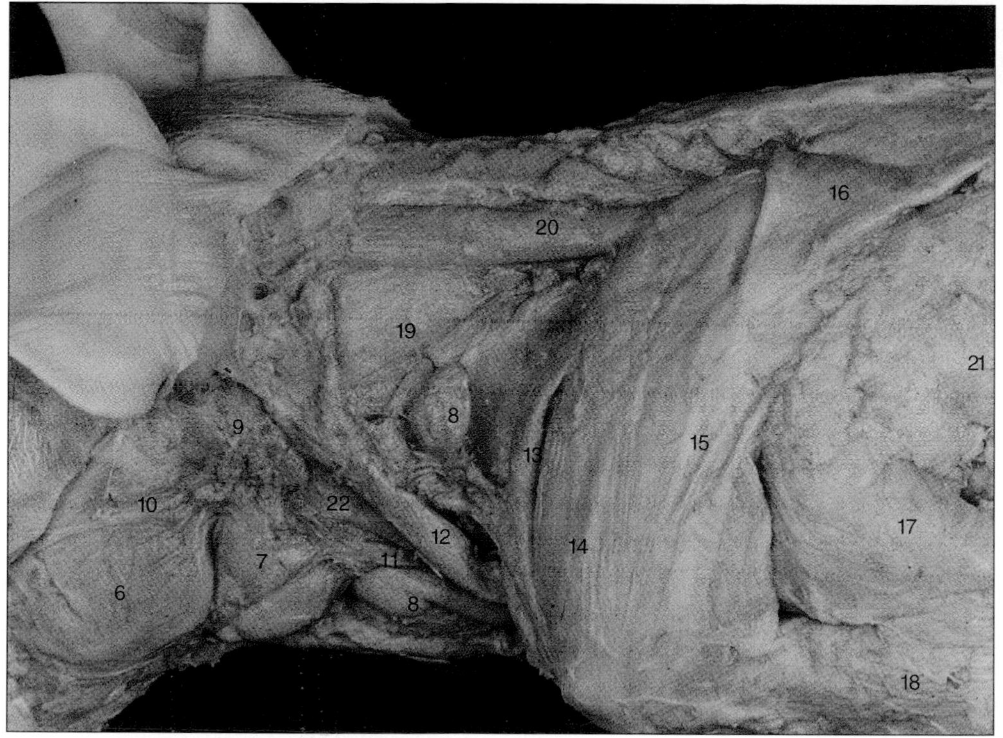

Figure 4-33
Superficial Structures of the Neck, Shoulder, and Thoracic Limb of the Fetal Pig, Left Lateral View

1. Brachioradialis muscle
2. Extensor carpi radialis muscle
3. Extensor digitorum communis muscle
4. Extensor digitorum lateralis muscle
5. Extensor carpi ulnaris muscle
6. Masseter muscle
7. Submandibular gland
8. Lymph node
9. Parotid gland
10. Salivary duct
11. External jugular vein
12. Clavotrapezius muscle
13. Clavobrachialis muscle
14. Acromiodeltoid muscle
15. Spinodeltoid muscle
16. Spinotrapezius muscle (cut)
17. Triceps brachii muscle (long head)
18. Triceps brachii muscle (lateral head)
19. Splenius capitis muscle
20. Rhomboideus capitis muscle
21. Latissimus dorsi muscle
22. Sternomastoid muscle

Figure 4-34
Superficial Muscles of the Hind Limb of the Fetal Pig, Left Lateral View

1. Lumbodorsal fascia
2. External oblique muscle (reflected)
3. Internal oblique muscle
4. Tensor fascia latae muscle (split)
5. Vastus lateralis muscle (under pin)
6. Gluteus medius muscle
7. Gluteus maximus muscle
8. Biceps femoris muscle
9. Semitendinosus muscle
10. Semimembranosus muscle
11. Testis
12. Gastrocnemius muscle
13. Soleus muscle
14. Achilles tendon
15. Flexor hallucis longus muscle
16. Tibialis anterior muscle

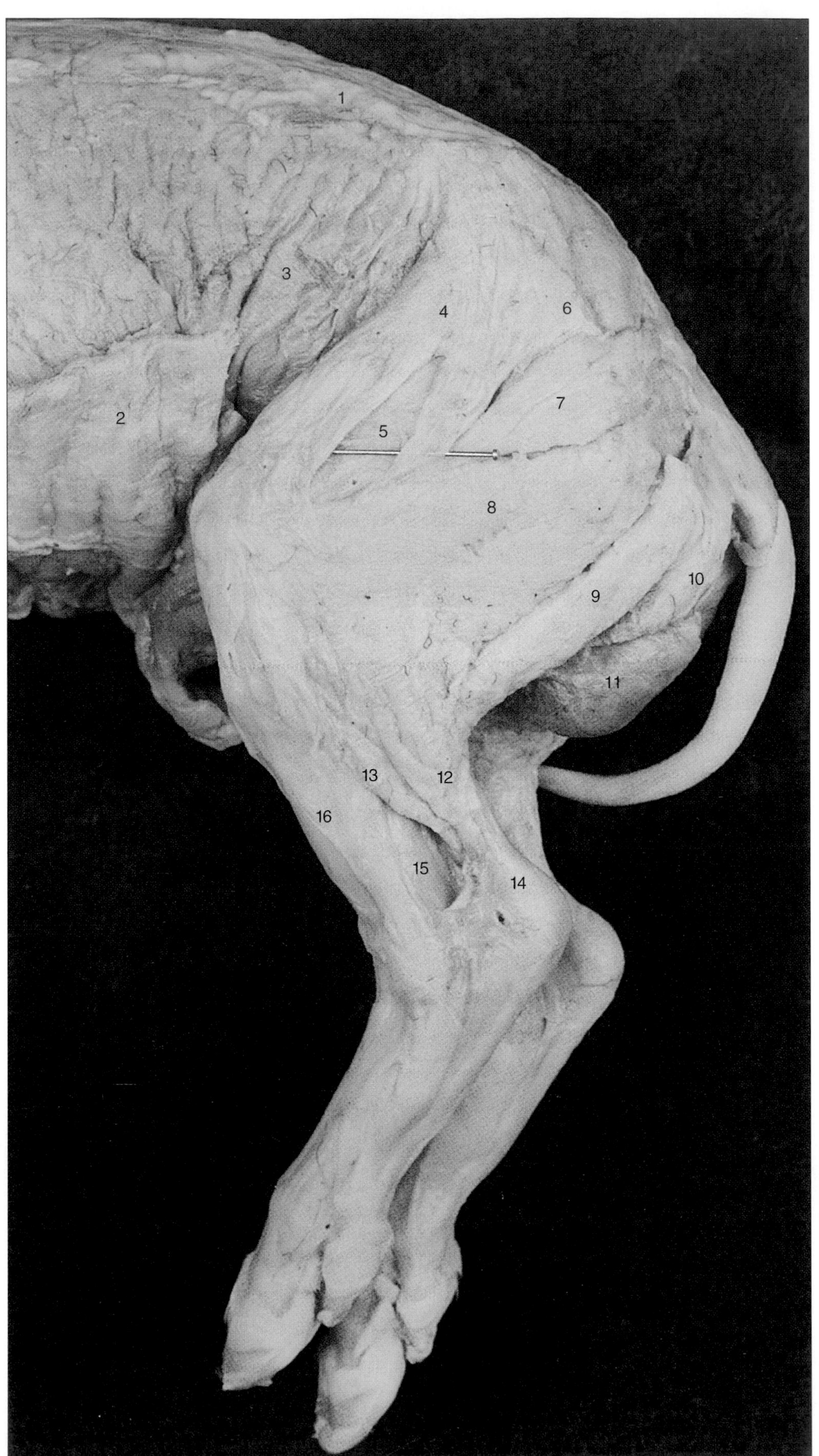

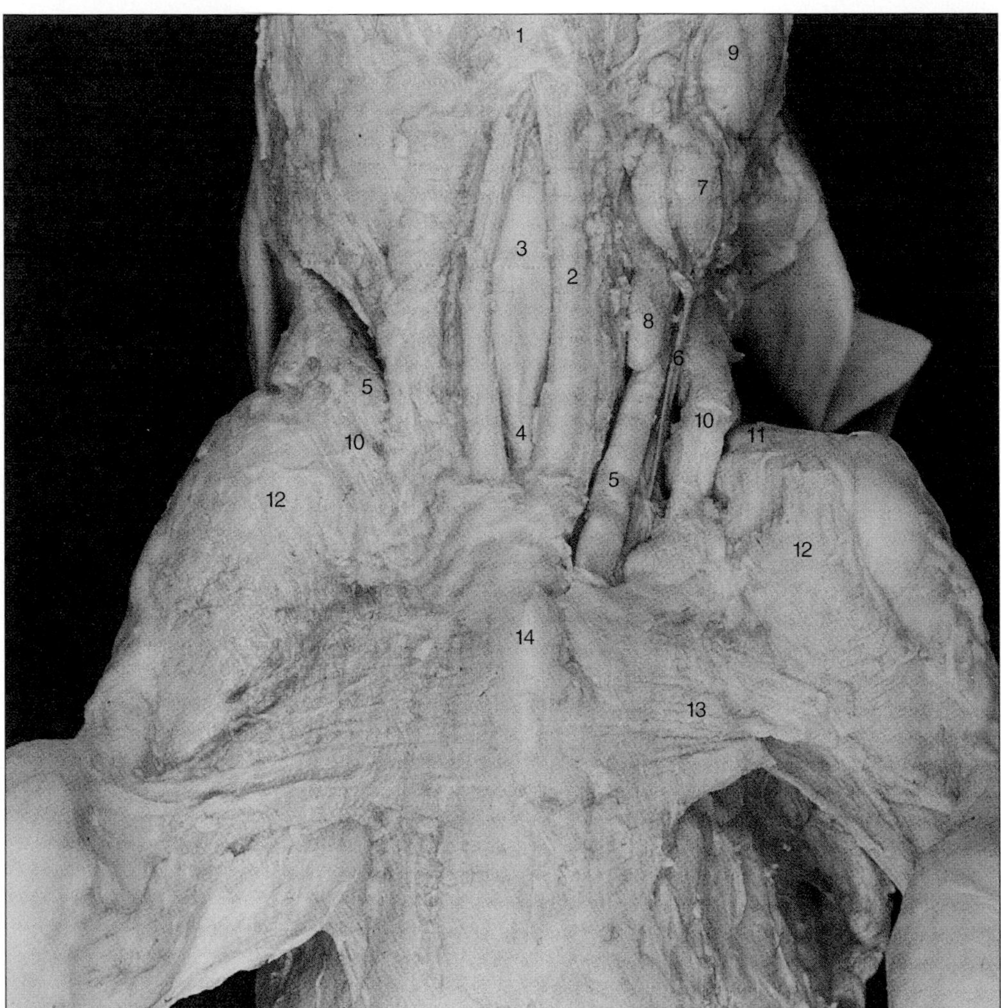

Figure 4-35
Superficial Anatomy of the Fetal Pig Neck and Shoulders, Ventral View

1. Mylohyoid muscle
2. Sternohyoid muscle
3. Larynx
4. Trachea
5. Sternomastoid muscle
6. External jugular vein
7. Lymph node
8. Submandibular gland
9. Masseter muscle
10. Clavotrapezius muscle
11. Acromiodeltoid muscle
12. Clavobrachialis muscle
13. Pectoralis major muscle
14. Sternum

Figure 4-36
Deep Anatomy of the Fetal Pig, Neck and Thorax

1. Larynx
2. Trachea
3. Thyroid gland
4. Common carotid artery
5. Vagus nerve
6. Internal jugular vein
7. External jugular vein (pinned bilaterally)
8. Cephalic vein
9. Subclavian vein
10. Superior vena cava
11. Internal mammary vein (cut, laid on lung tissue)
12. Right auricle
13. Left auricle
14. Heart
15. Lung
16. Diaphragm

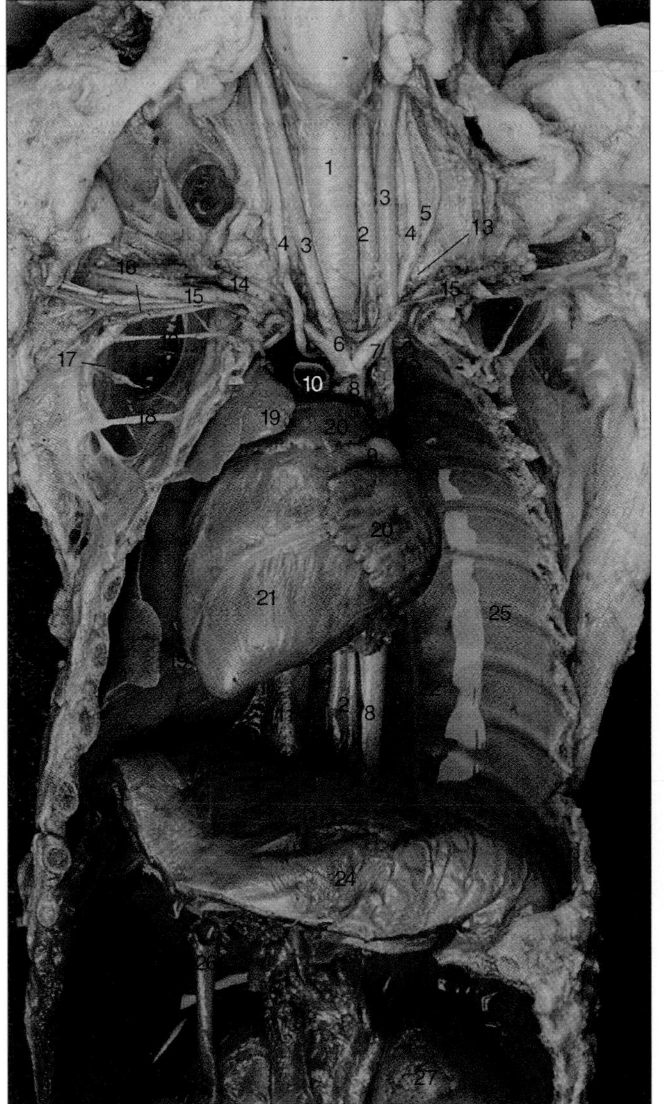

A

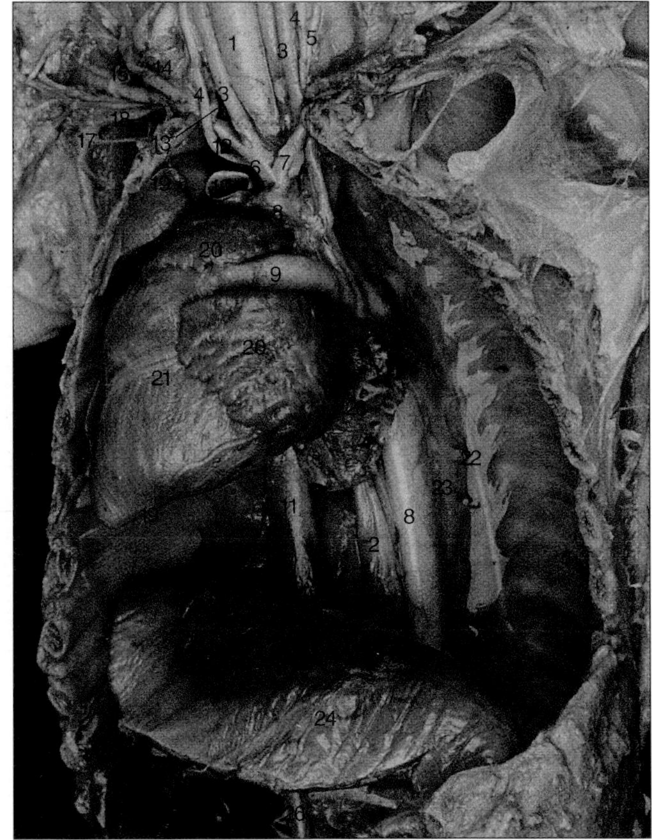

B

Figure 4-37
Arteries of the Neck and Thorax of the Fetal Pig, Left Lung Removed
(A) Heart in Normal Position
(B) Heart Reflected to Right

1. Trachea
2. Esophagus
3. Common carotid artery
4. Vagus nerve
5. Sympathetic trunk
6. Right brachiocephalic artery
7. Left brachiocephalic artery
8. Aorta
9. Ductus arteriosus
10. Superior vena cava (cut)
11. Inferior vena cava
12. Subclavian artery
13. Vertebral artery
14. Transverse scapular artery
15. Axillary artery
16. Radial nerve
17. Thoracodorsal nerve
18. Dorsal rami of thoracic nerves
19. Lung
20. Right and left auricles
21. Heart
22. Continuation of sympathetic trunk
23. Azygous vein
24. Diaphragm
25. Rib with costal artery and vein
26. Ductus venosus
27. Kidney
28. Hilum of left lung (with bronchi and blood vessels cut)

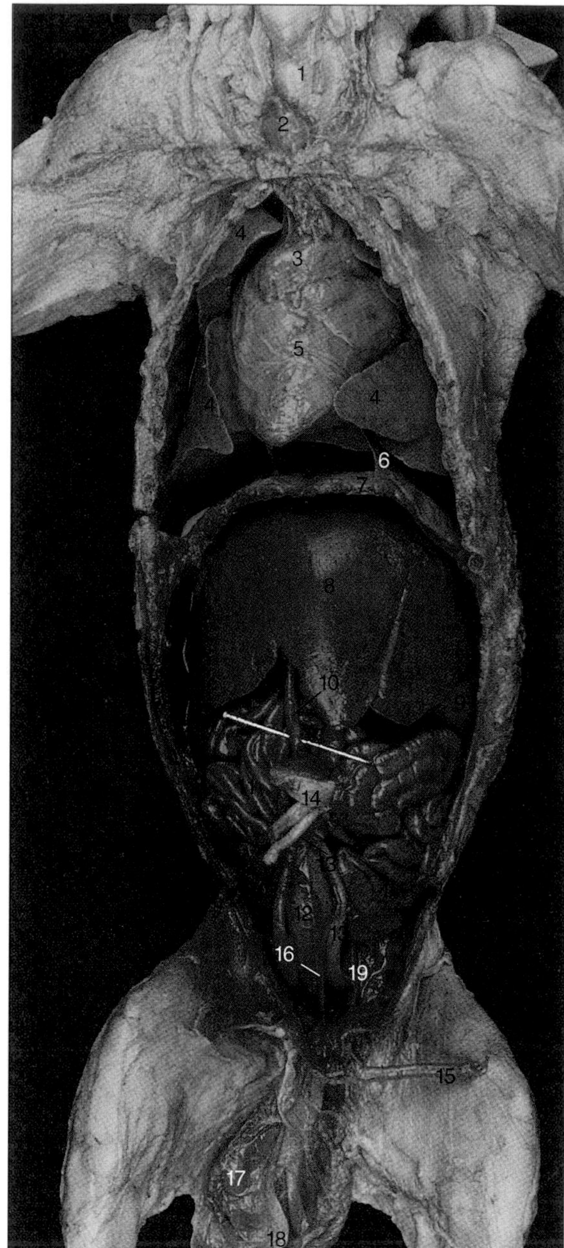

Figure 4-38
Thoracic and Abdominal Viscera of the Fetal Pig, Male

1. Trachea
2. Thyroid gland
3. Thymus
4. Lung
5. Heart in pericardium
6. Mediastinal membrane
7. Diaphragm
8. Liver
9. Spleen
10. Umbilical vein (on pin)
11. Small intestine
12. Urinary bladder
13. Umbilical arteries
14. Skin of umbilicus
15. Penis
16. Urethra
17. Testis
18. Epididymis
19. Spermatic cord (contains spermatic artery and vas deferens, which curves to pass behind base of urinary bladder)

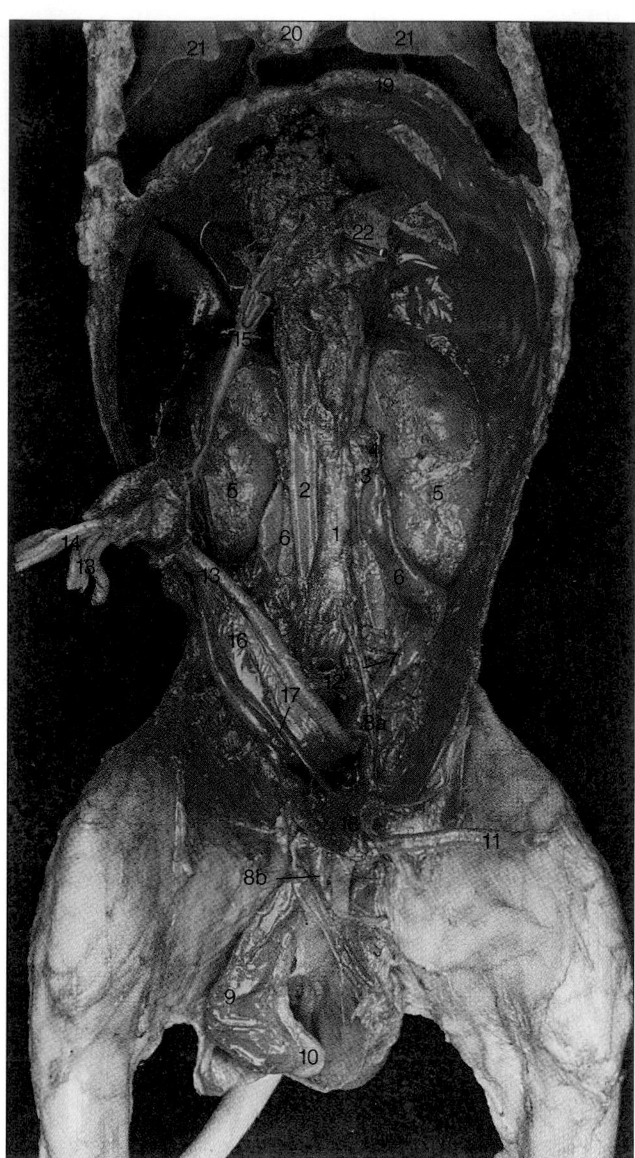

Figure 4-39
Abdominopelvic Cavity of the Fetal Pig, Male, Digestive Viscera Removed

1. Abdominal aorta
2. Abdominal vena cava
3. Renal artery
4. Renal vein
5. Kidney
6. Ureter
7. Spermatic artery
8. a. Vas deferens
   b. Vas deferens (in spermatic cord)
9. Testis
10. Epididymis
11. Penis
12. Rectum (cut)
13. Umbilical arteries
14. Umbilical vein
15. Ductus venosus (only remnants of liver remain)
16. Urinary bladder
17. Urethra
18. Prostate gland
19. Diaphragm
20. Heart
21. Lung
22. Pylorus of stomach (pin in antrum)

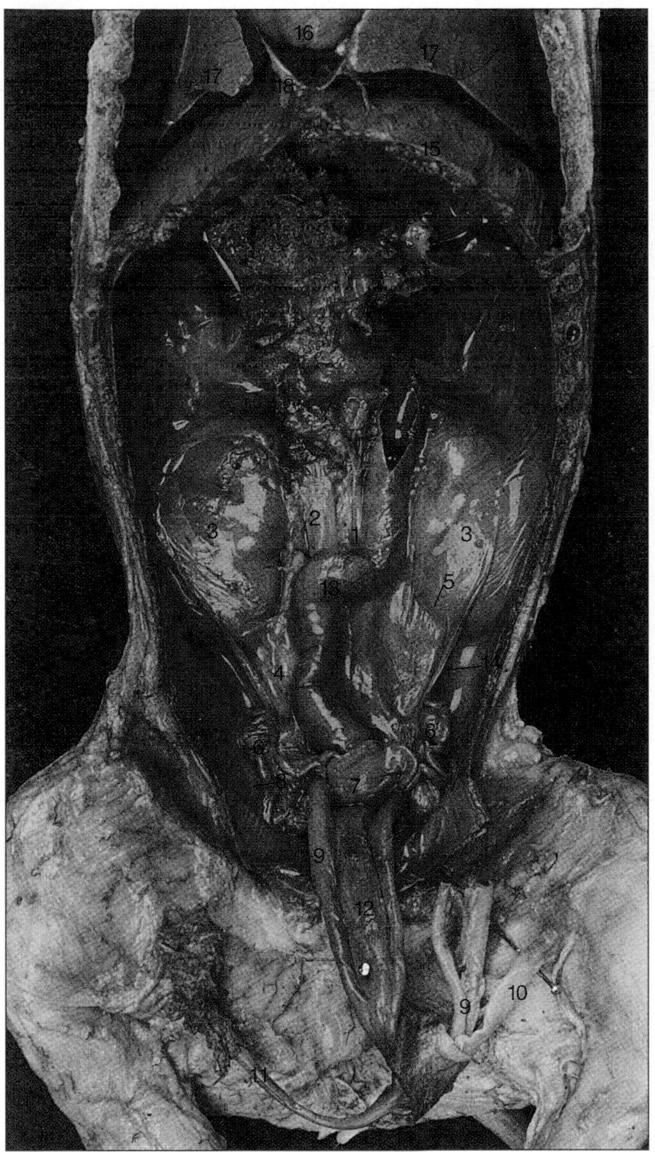

Figure 4-40
Abdominopelvic Cavity of the Fetal Pig, Female, Digestive Viscera Removed

1. Abdominal aorta
2. Abdominal vena cava
3. Kidney (behind intact peritoneum)
4. Ureter (behind intact peritoneum)
5. Ovarian artery
6. Ovary
7. Uterus
8. Uterine horn
9. Umbilical arteries
10. Umbilical vein (lying on pin)
11. Ductus venosus
12. Urinary bladder (reflected and pinned)
13. Sigmoid colon
14. Suspensory ligament of ovary
15. Diaphragm
16. Heart
17. Lung
18. Mediastinal membrane

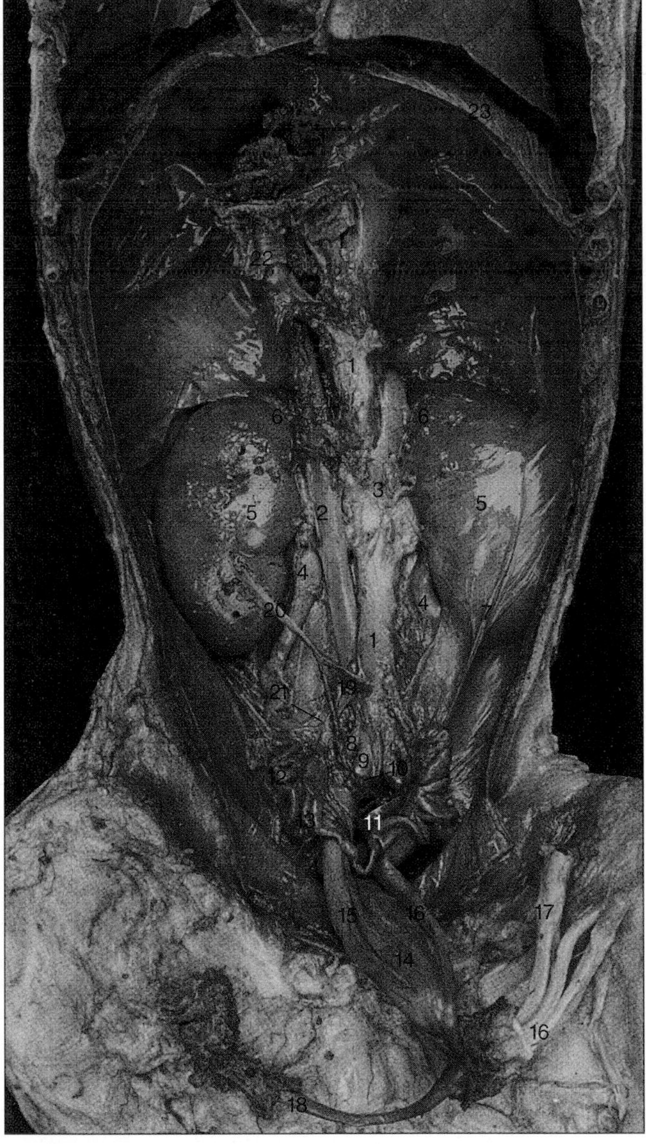

Figure 4-41
Deep Anatomy of the Abdominopelvic Cavity of the Fetal Pig, Abdominal Viscera Removed, Female

1. Abdominal aorta
2. Abdominal vena cava
3. Renal vein
4. Ureter
5. Kidney (left kidney behind peritoneum)
6. Adrenal gland
7. Suspensory ligament of ovary
8. External iliac artery
9. Internal iliac artery
10. Median sacral (caudal) artery
11. Rectum (cut)
12. Ovary
13. Uterine horn
14. Urinary bladder (reflected)
15. Urethra
16. Umbilical arteries
17. Umbilical vein
18. Ductus venosus
19. Posterior (inferior) mesenteric artery
20. Colic artery (pinned to kidney for clarity due to missing colon)
21. Superior hemorrhoidal artery
22. Remnant of small intestine (duodenum)
23. Diaphragm

Figure 4-42
General Anatomy of the Male Rat, Abdominal Cavity Exposed, Ventral View

1. Thorax
2. Abdomen
3. External oblique muscle (reflected and pinned)
4. Internal oblique muscle (lying on pin)
5. Transversus abdominis
6. Rectus abdominis
7. Peritoneum
8. Inferior epigastric artery
9. Liver
10. Spleen
11. Kidney
12. Rectum
13. Abdominal fat (small intestine not visible in this photograph)
14. Scrotum
15. Penis
16. Sternum (xiphoid process)

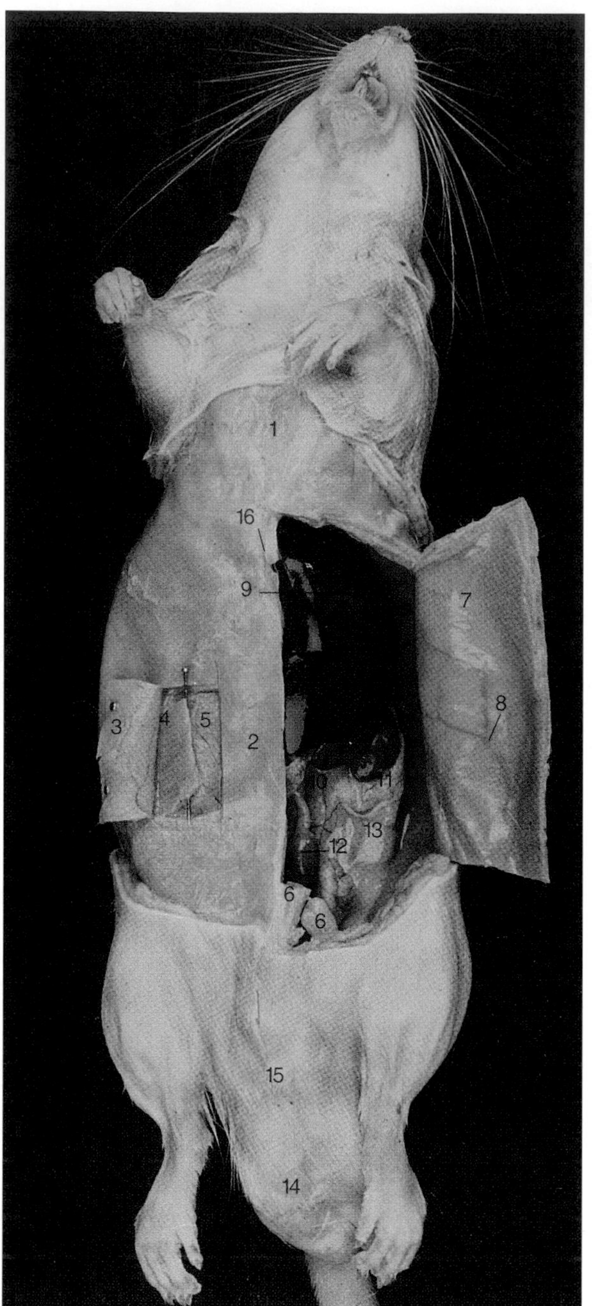

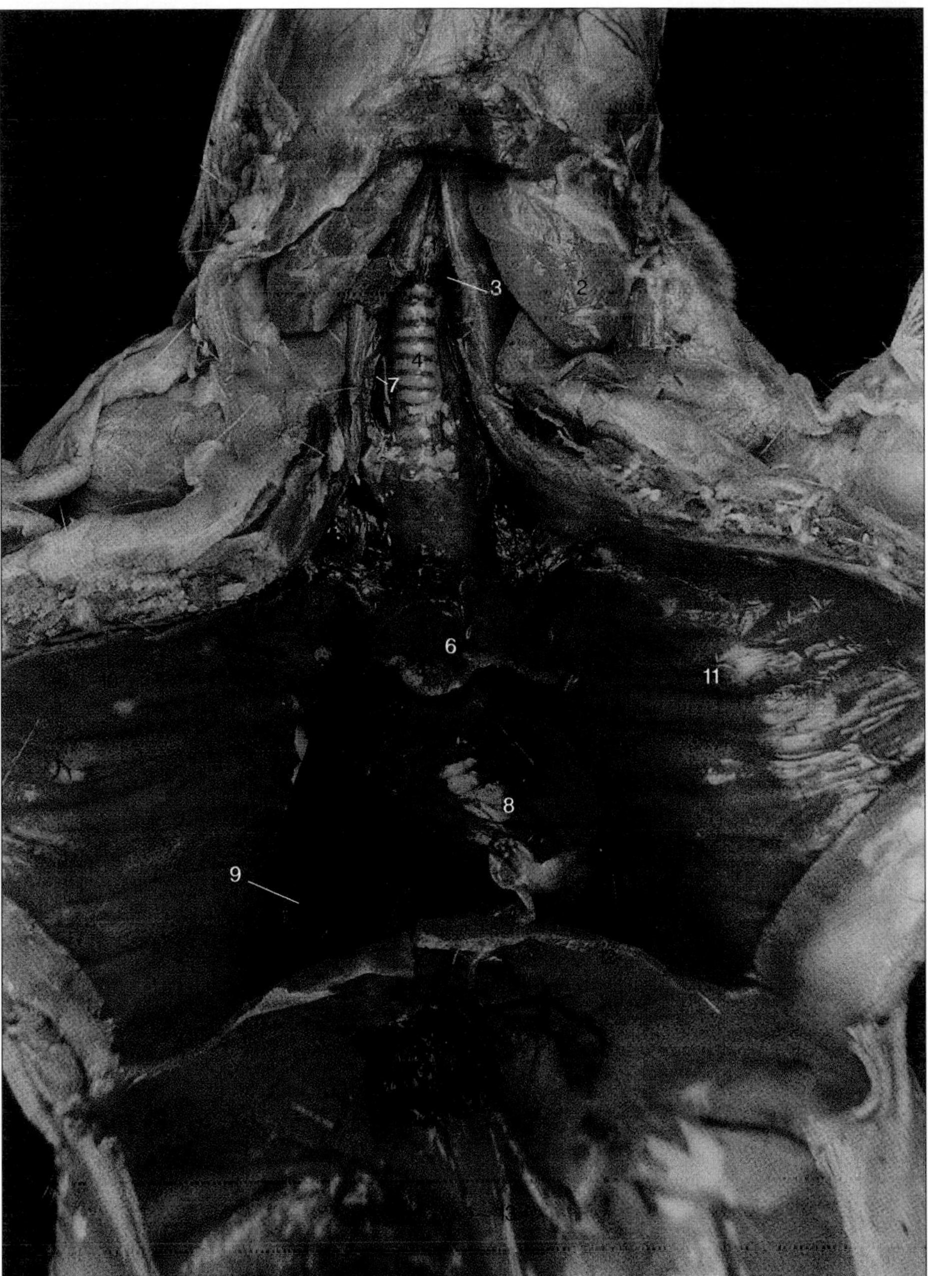

Figure 4-43
Deep Anatomy of the Rat, Neck and Thorax

1. Larynx
2. Salivary gland
3. Thyroid gland
4. Trachea
5. Sternohyoid muscle (unavoidably damaged on animal's right side during vascular perfusion)
6. Thymus
7. Common carotid artery
8. Heart
9. Lung
10. Internal mammary vein
11. Rib and intercostal artery and vein
12. Diaphragm
13. Crus of diaphragm

Figure 4-44
Abdominopelvic Cavity of the Male Rat

1. Sternum (xiphoid process)
2. Stomach
3. Liver
4. Small intestine (duodenum)
5. Pancreas
6. Spleen
7. Kidney
8. Small intestine (jejuneum and ileum)
9. Large intestine (cecum)
10. Rectum
11. Abdominal fat
12. Urinary bladder
13. Rectus abdominis muscle (cut)
14. Testis in scrotum
15. Epididymis
16. Penis
17. Seminal vesicle
18. Cremasteric fascia

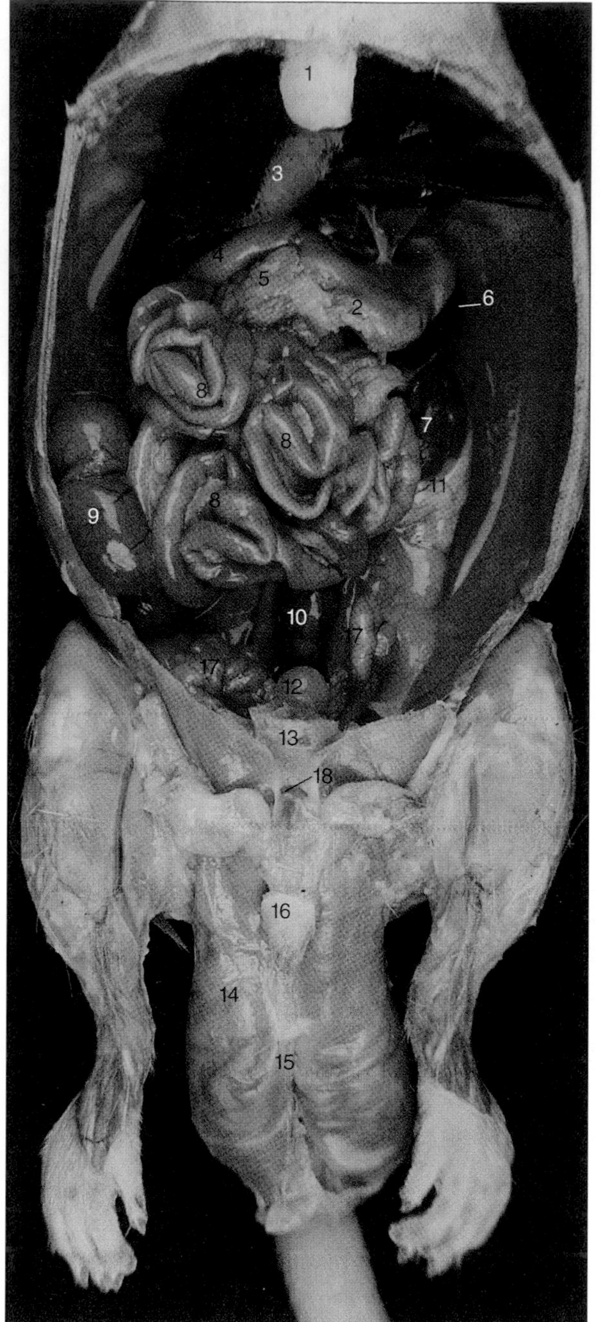

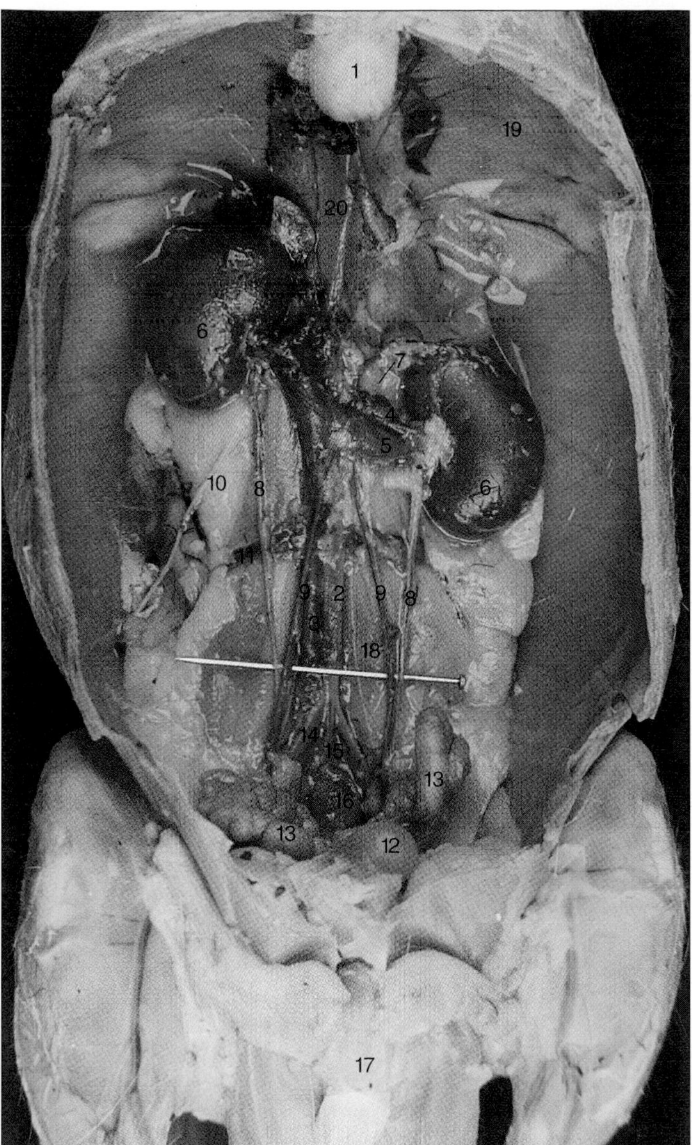

Figure 4-45
Abdominopelvic Cavity of the Rat, Male, Digestive Viscera Removed

1. Sternum (xiphoid process)
2. Abdominal aorta
3. Abdominal vena cava
4. Renal artery
5. Renal vein
6. Kidney
7. Adrenal gland
8. Ureter (lying on pin)
9. Spermatic artery (lying on pin)
10. Lumbar nerve (medial branch, extended for clarity)
11. Iliolumbar artery and vein
12. Urinary bladder
13. Seminal vesicle
14. Common iliac artery
15. Median sacral (caudal) artery
16. Rectum (cut)
17. Penis
18. Psoas major and minor muscles
19. Diaphragm
20. Crus of diaphragm

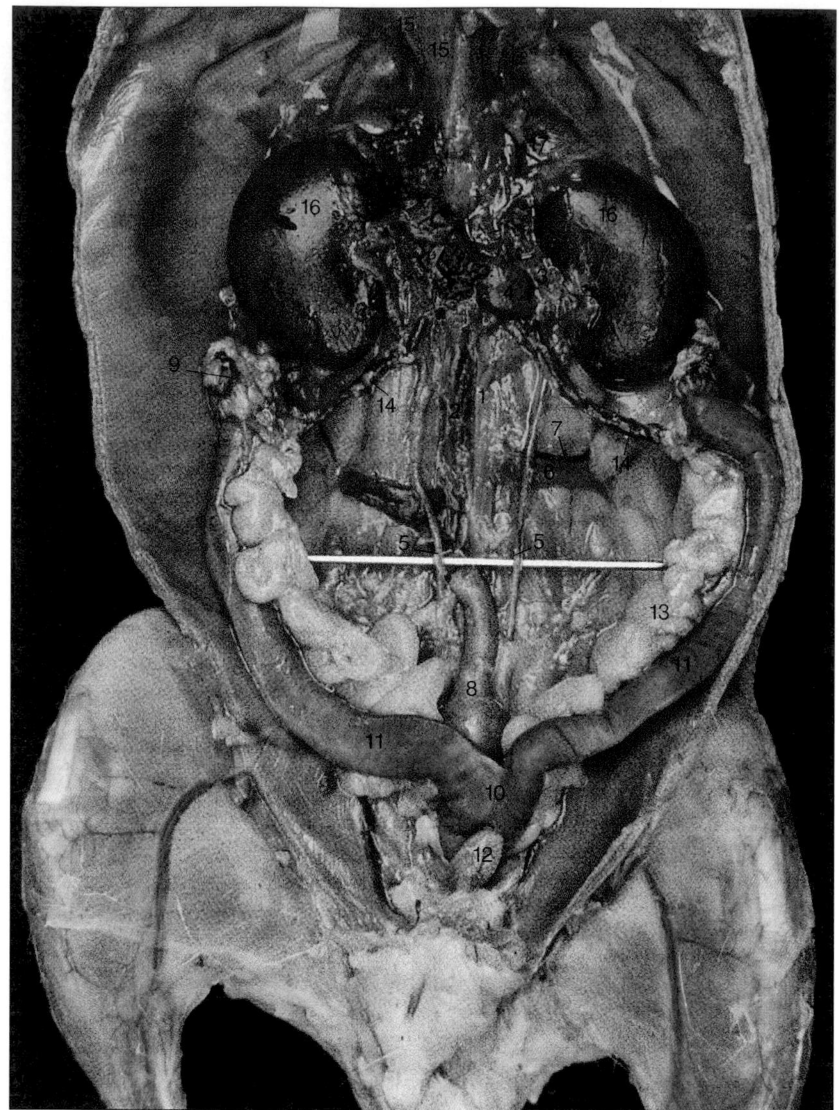

Figure 4-46
Abdominopelvic Cavity of the Rat, Female, Digestive Viscera Removed

1. Abdominal aorta
2. Abdominal vena cava
3. Renal artery
4. Renal vein
5. Ureter (lying on pin)
6. Iliolumbar artery
7. Iliolumbar vein
8. Rectum (cut)
9. Ovary
10. Uterus
11. Uterine horn
12. Urinary bladder
13. Abdominal fat
14. Ovarian artery and vein
15. Crus of diaphragm
16. Kidney
17. Adrenal gland

C H A P T E R 5

Reference Tables

Nerve-muscle connection. Light micrograph of neuromuscular synapses.

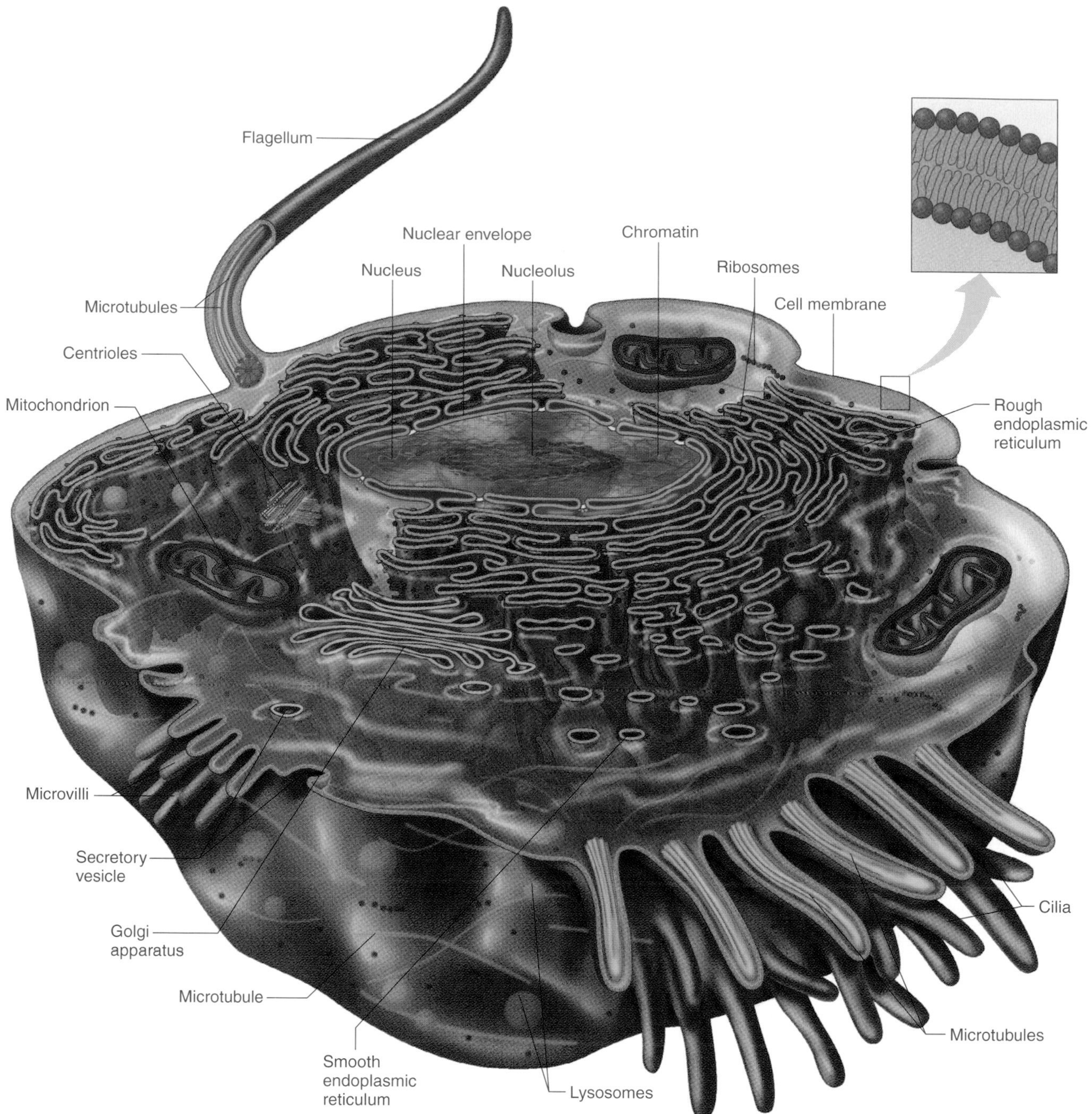

Figure 5-1
Generalized Illustration of a Cell

TABLE 5.1 Structure and function of some cellular components

STRUCTURE	DESCRIPTION AND FUNCTION
MEMBRANOUS	
Plasma membrane	Composed mainly of phospholipid bilayer with globular proteins floating dynamically on, in, and through it. Separates living cell contents from nonliving environment. Maintains cellular integrity. Embedded molecules serve as identifying cell markers (antigens), receptor molecules for hormones and related substances, signal transducers, selective ion channels, and transporter mechanisms.
Endoplasmic reticulum	Complex of membranous canals, sacs, and vesicles. Transports material within the cell; provides attachment for ribosomes; contributes to synthesis of lipids, steroids, and some carbohydrates used to form glycoproteins.
Golgi apparatus	Flattened membranous sacs. Synthesizes and packages carbohydrates and glycoproteins.
Lysosomes	Small membranous sacs. Contains enzymes used in intracellular digestion.
Peroxisomes	Small membranous vesicles. Contains peroxidase enzymes used in breakdown of complex toxins and other organic molecules.
Mitochondria	Small membranous sacs with complex internal structure and separate DNA. Contains enzymes of Krebs cycle; central to carbohydrate metabolism and synthesis of ATP.
Nucleus	Nuclear contents, notably DNA, separated from cytoplasm by porous nuclear envelope.
NONMEMBRANOUS	
Ribosomes	Small structures composed of two parts containing protein and RNA molecules. Often associated with endoplasmic reticulum. Synthesizes proteins under instructions of messenger RNA triplet code.
Centrosome	Double structure composed of two, short, rod-like centrioles. Important in distribution of chromosomes during cell division and in formation of cilia.
Microfilaments and microtubules	Composed of protein complexes. Acts as cytoskeletal framework. Functions in whole-cell and local membrane movements, cellular elasticity, and formation of cellular extensions (e.g., microvilli).
Cilia and flagella	Movable membranous extensions. Important in movement of fluid environment over stationary cell surface (cilia) and cell itself (flagellum of sperm cell).
Nucleolus	Dense object composed of protein and RNA molecules. Essential in ribosome formation.

TABLE 5.2 Some membrane transport processes

PROCESS	DESCRIPTION
PHYSICAL PROCESSES: DO NOT REQUIRE LOCAL EXPENDITURE OF METABOLIC ENERGY	
Bulk flow	Movement of substances from higher pressures toward lower pressures. Examples: movement of gases in and out of ventilatory tree during breathing, movement of blood through arteries and veins due to pumping action of heart.
Diffusion	Movement of ions or molecules from higher concentrations toward lower concentrations due to random molecular collisions. Examples: movement of sodium and potassium ions and glucose molecules in extracellular fluid.
Filtration	Bulk flow through a semipermeable membrane. Example: movement of fluid and small molecules through kidney capillary walls due to hydrostatic pressure.
Dialysis	Diffusion of solute molecules through a semipermeable membrane. Example: passage of lipid-soluble substances, such as steroid molecules, through cell membrane.
Osmosis	Diffusion of water down *its* concentration gradient through a semipermeable membrane. Osmosis generally operates *against* concentration gradient of solute(s) to which the membrane is *im*permeable. Example: net movement of extracellular fluid into the venous ends of capillaries under influence of *im*permeant plasma proteins.
Facilitated diffusion	Diffusion through an otherwise impermeable membrane by means of carrier molecules. Example: movement of glucose through muscle cell membranes (requires insulin to enhance action of facilitating carriers).
PHYSIOLOGICAL PROCESSES: REQUIRE LOCAL EXPENDITURE OF METABOLIC ENERGY	
Active transport	Carrier-mediated transport of ions or molecules through a living membrane via energy-requiring shape change of carrier molecule. Energy expenditure permits transport from lower to higher concentration. Examples: movement of sodium from inside to outside of resting nerve cells; transport of potassium and calcium from outside to inside cells, thereby causing high internal concentrations of these ions.
Phagocytosis and pinocytosis	Transport of large particles or fluid into a cell via engulfing action of membrane followed by pinching off to form an intracellular vesicle. Both are processes of endocytosis. Example: trapping of bacteria by white blood cells.
Exocytosis	Transport of substances out of a cell by fusion of internal vesicle with cell membrane and release of contents to the exterior. Examples: secretion of hormones and neurotransmitters, such as prolactin and acetylcholine.

TABLE 5.3 Formed elements of blood

CELL TYPE	DESCRIPTION (WRIGHT'S STAIN)	NORMAL NUMBER (CELLS/μL OF BLOOD)	FUNCTION
Erythrocytes (Red blood cells, RBC)	7.5μ diameter, biconcave disk, no nucleus	4–6 million	Transport of respiratory gases (O_2 and CO_2)
Leukocytes (White blood cells, WBC)		5,000 to 10,000/mm^3	Aid in defense against infections by microorganisms
Granulocytes			
Neutrophil	12–15μ diameter, multilobed nucleus, small pink-purple granules	3,000–7,000 (65% of total leukocytes)	Phagocytosis; elevated in number during acute infections
Eosinophil	10–14μ diameter, bilobed nucleus, large orange granules	100–400 (3% of total leukocytes)	Destroys antigen-antibody complexes, phagocytosizes parasites, involved in allergic response
Basophil	8–12μ diameter, bilobed large purple granules that may obscure nucleus	20–50 (1% of total leukocytes)	Contains biogenic amines; releases heparin, histamine, other chemicals during inflammatory response
Agranulocytes			
Lymphocyte	5–16μ diameter, round or nucleus, indented, single-lobed nucleus, variable amount of cytoplasm	1,500–3,000 (25% of total leukocytes)	Immune response by direct cellular contact or via antibody production; elevated in infectious mononucleosis; suppressed by steroid therapy
Monocyte	12–20μ diameter, horseshoe-shaped nucleus	100–700 (6% of total leukocytes)	Macrophages; phagocytosis
Platelets	2-4μ, appear as cytoplasmic fragments	25,000 to 500,000	Coagulation

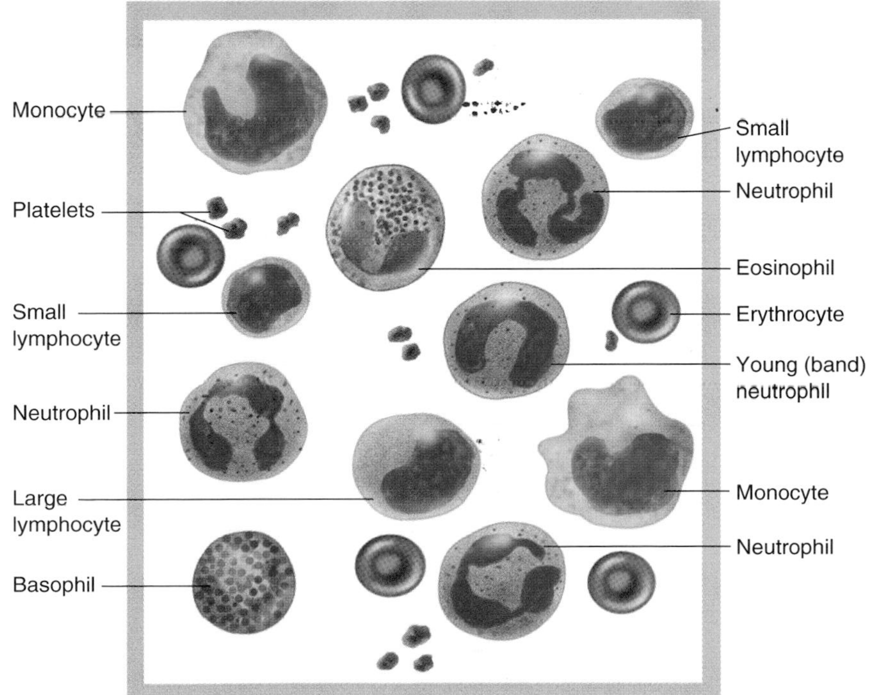

Figure 5-2
The Formed Elements of Blood.
The structure of red blood cells, white blood cells, and platelets.

TABLE 5.4 Terms for bone structure

TERM	DEFINITION
Epiphysis	Either rounded end of head of a long bone
Diaphysis	The shaft of a long bone
Anatomic neck	The epiphyseal growth plate
Surgical neck	The narrow part of a long bone, just past the head, where fracture is most likely
Ramus	A branch
Cornu	A horn
Hamulus	A hook
Lingula	A tongue
Foramen (pl. foramina)	A hole; an opening into or through a bone to permit passage of blood vessels, nerves, or ligaments
Fossa	A valley; a relatively deep pit or depression
Fovea	A relatively small pit or depression
Sulcus	A narrow valley
Meatus	A tunnel
Trochanter	A large, blunt, rounded process that serves as a site for muscle attachment
Tubercle	A small, blunt, rounded process that serves as a site for muscle attachment
Tuberosity	A large, rounded, often rough eminence or surface that serves as a site for muscle attachment
Condyle	A large, rounded process at the end of a bone, usually contributing to a joint
Epicondyle	A smaller, rounded process at the end of a bone, on top of a condyle, usually contributing to a joint
Trochlea	A pulley; a smooth notched surface often found at a joint
Facet	A face; a smooth, nearly flat surface at a joint
Fissure	A crack or cleft
Crest or crista	A narrow ridge
Spine	A pointed ridge
Fontanel	Specifically, six spaces between the cranial bones of the fetal and infant skull prior to closure of the sutures
Second and fifth intercostal spaces	Specifically refers to a place between the 2nd and 3rd rib and a place between the 5th and 6th ribs where the second and first heart sounds, respectively, can be heard especially well

TABLE 5.5 Bones of the human skeleton

PART OF THE BODY	NAMES OF BONES
AXIAL SKELETON (80 BONES TOTAL)	
Skull (28 bones)	
Cranium (8 bones)*	Frontal (1) Parietal (1 pair) Temporal (1 pair) Occipital (1) Sphenoid (1) Ethmoid (1)
Face (14 bones)	Lacrimal (1 pair) Nasal (1 pair) Palatine (1 pair) Inferior nasal conchae (1 pair) Vomer (1) Maxillae (1 pair) Zygomatic (1 pair) Mandible (1)
Middle ear (6 bones)	Malleus (1 pair) Incus (1 pair) Stapes (1 pair)
Hyoid bone (1)	
Spinal column (26 bones total)	Cervical vertebrae (7) Thoracic vertebrae (12) Lumbar vertebrae (5) Sacrum (4–5 fused into 1) Coccyx (4–5 fused into 1)
Sternum and ribs (25 bones total)	Sternum (1) True ribs (7 pairs) False ribs (5 pairs)
APPENDICULAR SKELETON (126 BONES TOTAL)	
Shoulder girdle and arm (64 bones total)	Clavicle (1 pair) Scapula (1 pair) Humerus (1 pair) Ulna (1 pair) Radius (1 pair) Carpals (8 pairs: scaphoid, lunate, triquetum, pisiform, trapezium, trapezoid, capitate, hamate) Metacarpals (5 pairs) Phalanges (14 pairs)
Pelvic girdle and leg (62 bones total)	Os coxae (1 pair: 2 pelvic bones each formed by fusion of ilium, ischium, and pubis) Femur (1 pair) Patella* (1 pair) Tibia (1 pair) Fibula (1 pair) Tarsals (7 pairs: Talus, calcaneus, navicular, medial cuneiform, intermediate cuneiform, lateral cuneiform, cuboid) Metatarsals (5 pairs) Phalanges (14 pairs)

*A variable number of rounded bones known as **sesamoid bones** (because of their supposed resemblance to sesame seeds) may appear in various tendons, especially those in the wrist, knee, ankle, and foot. Only two of them, the patellae, are commonly found. **Wormian bones** are found in variable numbers within the suture lines of the skull. While most are commonly smaller than the size of fingernails, some can be surprisingly large.

TABLE 5.6 Comparison of female and male skeletons

Differences between male and female skeletons are graded, not discrete. Female skeletons can have many masculine features, and vice versa. Nevertheless, there are trends, including those listed below. A typically masculine pelvis is called *android;* a typically feminine pelvis is called *gynecoid.* Many intermediate types exist.

PORTION OF SKELETON	FEMALE	MALE
GENERAL FORM	Bones lighter and thinner Muscle attachment sites smaller and smoother Joint surfaces relatively small	Bones heavier and thicker Muscle attachment sites larger and rougher Joint surfaces relatively large
PELVIS		
Pelvic cavity	Wider in all dimensions Shorter and roomier Pelvic outlet relatively large	Smaller in all dimensions Deeper Pelvic outlet usually obstructed
Sacrum	Short, wide, flat concavity more pronounced in a posterior direction; sacral promontory less pronounced	Long, narrow, with smooth concavity of sacral curvature; sacral promontory more pronounced
Coccyx	More movable and follows posterior direction of sacral curvature	Less movable
Pubic arch	Greater than a 90° angle	Less than a 90° angle
Ischial spine, ischial tuberosity, and anterior super iliac spine	Oriented outward and further apart	Oriented inward
Greater sciatic notch	Narrow	Wide

TABLE 5.7 Extrinsic muscles of the eye

NAME	ORIGIN	INSERTION	ACTION	INNERVATION
Rectus superior	Tendinous ring of tissue which surrounds optic foramen at back of orbit	Top of eyeball	Rolls eye upward	Oculomotor
Rectus inferior		Bottom of eyeball	Rolls eye downward	Oculomotor
Rectus lateralis		Lateral side of eyeball	Rolls eye laterally	Abducens
Rectus medialis		Medial side of eyeball	Rolls eye medially	Oculomotor
Obliquus superior		Top of eyeball under rectus superior, through trochlea	Prevents rotation of eyeball on axis; directs gaze down and laterally	Trochlear
Obliquus inferior	Maxilla at front of orbit	Lateral side of eyeball under rectus lateralis	Prevents rotation of eye on axis; directs gaze up and laterally	Oculomotor

TABLE 5.8 Facial muscles

NAME	ORIGIN	INSERTION	ACTION	INNERVATION
Buccinator	Maxillary and mandibular alveolar processes	Into orbicularis oris at sides of mouth	Compresses cheek, retracts corner of mouth as in playing a brass musical instrument	Facial
Orbicularis oris	Maxillae, mandible, nasal septum	Fibers encircle mouth, insert on fascia	Puckering, shaping of mouth in speech	Facial
Orbicularis oculi	Maxillae, frontal bone	Fibers encircle orbit	Closes eye, assists in squinting	Facial
Occipitofrontalis	Occipital bone	Skin around eyebrows and above nose	Moves scalp, elevates eyebrows	Facial
Zygomaticus major	Zygomatic bone	Into orbicularis oris at corners of mouth	Retracts and elevates corners of mouth as in smiling	Facial
Zygomaticus minor	Zygomatic bone	Into orbicularis oris of upper lip	Elevates upper lip, assists in smiling	Facial
Levator palpebrae superioris	Lesser wing of sphenoid	Skin of upper eyelid	Elevates upper eyelid	Oculomotor
Corrugator supercilii	Bridge of nose, orbicularis oculi	Skin of eyebrows	Depresses and adducts eyebrows; furrows forehead as in frowning	Facial

TABLE 5.9 Chewing muscles

NAME	ORIGIN	INSERTION	ACTION	INNERVATION
Masseter	Zygomatic arch and maxilla	Lateral surface of mandible	Closes jaw	Trigeminal
Temporalis	Temporal bone	Coronoid process of mandible	Closes jaw	Trigeminal
Pterygoid (medial and lateral)	Pterygoid processes of sphenoid bone	Medial surface of mandible	Moves jaw from side to side; grates teeth for chewing	Trigeminal

TABLE 5.10 Muscles of the throat

NAME	ORIGIN	INSERTION	ACTION	INNERVATION
Digastric	Mastoid process of temporal bone	Mandible	Elevates hyoid bone; depresses and retracts mandible	Posterior portion: Facial Anterior portion: Mandibular branch of trigeminal
Mylohyoid	Mandible	Hyoid	Elevates floor of mouth when mandible is fixed; depresses mandible when hyoid is fixed	Mandibular division of trigeminal
Omohyoid	Superior border of scapula and tendon from clavicle	Hyoid	Depresses hyoid; stabilizes hyoid when opening mouth	C1–C3 via ansa hypoglossi
Sternohyoid	Manubrium of sternum; costal cartilage 1	Hyoid	Depresses hyoid; stabilizes hyoid when opening mouth	C1–C3 via ansa hypoglossi
Sternothyroid	Manubrium of sternum; costal cartilages 1 and 2	Thyroid cartilage	Depresses larynx; stabilizes larynx when opening mouth	Upper cervical nerves via ansa cervicalis and ansa hypoglossi

TABLE 5.11 Muscles of the tongue

NAME	ORIGIN	INSERTION	ACTION	INNERVATION
Intrinsic muscles: Longitudinal, vertical, and transverse	Within tongue	Within tongue	Change shape of tongue in speaking, chewing, licking	Hypoglossal
Genioglossus	Genu of mandible	Tongue	Depresses and protrudes tongue	Hypoglossal
Hyoglossus	Hyoid	Side of tongue	Depresses and retracts tongue	Hypoglossal
Styloglossus	Styloid process of temporal bone	Inferior and lateral aspect of tongue	Retracts tongue	Hypoglossal

NOTE: The three above-named muscles are **extrinsic muscles of the tongue,** so identified because their origins lie outside the muscular tongue itself.

TABLE 5.12 Muscles of the pharynx and palate

NAME	ORIGIN	INSERTION	ACTION	INNERVATION
Constrictor pharyngis inferior	Cricoid and thyroid cartilages	Median raphe of pharynx	Constricts lower pharynx during swallowing	Glossopharyngeal and vagus
Constrictor pharyngis medius	Greater and lesser cornu of hyoid	Median raphe of pharynx	Constricts middle pharynx during swallowing	Glossopharyngeal and vagus
Constrictor pharyngis superior	Middle pterygoid plate, mandible, floor of mouth	Median raphe of pharynx	Constricts upper pharynx during swallowing	Glossopharyngeal and vagus
Stylopharyngeus	Styloid process of temporal bone	Sides of pharynx; thyroid cartilage	Elevates and dilates pharynx	Glossopharyngeal
Palatopharyngeus	Soft palate	Pharynx	Narrows fauces; depresses palate; elevates pharynx	Glossopharyngeal and vagus
Palatoglossus	Soft palate	Tongue	Narrows fauces; elevates back of tongue	Pharyngeal plexus
Levator veli palatini	Temporal bone and cartilage of Eustachian tube	Soft palate	Elevates soft palate	Glossopharyngeal and vagus
Tensor veli palatini	Sphenoid bone and cartilage of Eustachian tube	Soft palate	Increases tension of soft palate; opens Eustachian tube as in yawning	Mandibular division of trigeminal

NOTE: The palatopharyngeus muscle and its mucous membrane covering form the clearly seen arch of the soft palate, from which hangs the uvula. Just anterior to this arch on each side is the palatoglossus muscle which, with its mucous membrane covering, forms the more lateral and less clearly seen glossopalatine arch. Between these two arches on each side is a fossa that houses the lymph node known as the palatine tonsil.

TABLE 5.13 Muscles that move the head

NAME	ORIGIN	INSERTION	ACTION	INNERVATION
Sternocleidomastoid	Sternum and clavicle	Mastoid process of temporal bone	Bows head, rotates head	Spinal accessory, C2–C3
Trapezius	Acromial process of clavicle and spine of scapula	Occipital bone, ligamentum nuchae, spines of 7th cervical and all thoracic vertebrae	Extends head, rotates head	Spinal accessory, C3–C4
Obliquus capitis inferior	Spinous process of axis	Transverse process of atlas	Rotates head	Branch of suboccipital
Splenius capitis	Ligamentum nuchae, spines of 7th cervical and top four thoracic vertebrae	Occipital bone and mastoid process of temporal bone	Extends head, rotates head	Middle and lower cervical spinal nerves
Semispinalis capitis	See MUSCLES OF THE VERTEBRAL COLUMN. The capitis division of this muscle inserts on the occipital bone. When the vertebrae serve as the origin and the occipital bone as the insertion, this muscle (bilaterally) extends the head or (unilaterally) draws the head toward the contracting side.			
Longissimus capitis	See MUSCLES OF THE VERTEBRAL COLUMN. The capitis division of this muscle inserts on the mastoid process of the temporal bone. When the vertebrae serve as the origin and the occipital bone as the insertion, this muscle (bilaterally) extends the head or (unilaterally) draws and rotates the head toward the contracting side.			

TABLE 5.14 Muscles that move the shoulder

NAME	ORIGIN	INSERTION	ACTION	INNERVATION
Trapezius	See MUSCLES THAT MOVE THE HEAD. If origin and insertion are reversed, this muscle causes elevation of shoulders, as in shrugging, by elevating clavicle and scapula.			
Pectoralis minor	Outer surface of third, fourth, and fifth ribs	Coracoid process of scapula	Depresses shoulder, rotates scapula forward and down; can assist in elevating ribs	Long thoracic
Serratus anterior	Outer surface of upper eight or nine ribs	Ventral surface of vertebral border of scapula	Rotates scapula forward and toward thoracic wall; can assist in elevating ribs	Spinal accessory, C3–C4
Rhomboideus major	Spines of second to fifth thoracic vertebrae	Vertebral border of scapula	Adducts scapula, rotates slightly upward	Dorsal scapular
Rhomboideus minor	Spines of seventh cervical and first thoracic vertebrae	Vertebral border of scapula	Adducts scapula	Dorsal scapular

NOTE: The **triangle of auscultation** is formed at the caudal medial border of the scapula by the edges of the latissimus dorsi, trapezius, and rhomboideus muscles.

TABLE 5.15 Muscles that move the upper arm

NAME	ORIGIN	INSERTION	ACTION	INNERVATION
Pectoralis major	Clavicle, sternum, cartilages of second to sixth ribs	Crest and greater tubercle of humerus	Flexes and adducts arm	Anterior thoracic
Latissimus dorsi	Spinous processes of lower six thoracic and all lumbar vertebrae, sacral spine, iliac crest and lower four ribs	Intertubercular groove of humerus	Extends, adducts, rotates arm medially, draws shoulder down and back	Thoracodorsal
Deltoideus	Clavicle and acromion and spine of scapula	Deltoid tuberosity of humerus	Abducts arm	Axillary
Coracobrachialis	Coracoid process of scapula	Medial surface of humerus	Adducts arm; assists in flexion and medial rotation	Musculocutaneous
Teres major	Medial border of scapula	Just distal to lesser tubercle of humerus	Adducts, extends, rotates arm medially	Lower subscapular
Teres minor	Medial border of scapula	Greater tubercle of humerus	Rotates arm laterally	Axillary
Subscapularis	Subscapular fossa of scapula	Lesser tubercle of humerus	Extends and medially rotates arm	Subscapular C5, C6
Supraspinatus	Supraspinous fossa of scapula	Greater tubercle of humerus	Initiates abduction of arm	Suprascapular C5, C6
Infraspinatus	Infraspinous fossa of scapula	Greater tubercle of humerus	Extends and laterally rotates arm	Suprascapular C5, C6

NOTE: The **rotator cuff** is formed from the tendons of the last four muscles named above because together they form a cuff that binds the humerus into the shallow glenoid fossa. A rotator cuff injury involves damage to one or more of these muscles or their tendons.

NOTE: Alone, the deltoid cannot initiate the first 15° of abduction, which is a duty of the supraspinatus muscle and its innervation and which is separate from that of the deltoid. Differential assessment of peripheral nerve injury is possible by asking a patient to abduct the arm from anatomical position.

TABLE 5.16 Muscles that move the lower arm

NAME	ORIGIN	INSERTION	ACTION	INNERVATION
Biceps brachii	Long head: Tuberosity above glenoid cavity of scapula Short head: Coracoid process of scapula	Radial tuberosity	Flexes and supinates arm and forearm	Musculocutaneous
Brachialis	Anterior surface of distal humerus	Tuberosity and coronoid process of ulna	Flexes forearm	Musculocutaneous, radial, and median
Brachioradialis	Supracondyloid ridge of humerus	Proximal to styloid process of radius	Flexes forearm	Radial
Triceps brachii	Long head: Infraglenoid tuberosity of scapula Lateral head: Posterior surface of humerus above radial groove Medial head: Posterior surface of humerus below radial groove	Olecranon process of ulna	Extends forearm	Radial
Anconeus	Lateral epicondyle of humerus	Olecranon process and proximal one-fourth of ulna	Extends forearm	Radial
Pronator teres	Medial epicondyle of humerus, coronoid process of ulna	Middle third of lateral surface of radius	Pronates and flexes forearm	Median
Pronator quadratus	Distal shaft of ulna	Distal shaft of radius	Pronates forearm	Median
Supinator	Lateral epicondyle of humerus, proximal end of ulna	Proximal third of radius	Supinates forearm	Median

TABLE 5.17 Muscles that move the wrist and hand

NAME	ORIGIN	INSERTION	ACTION	INNERVATION
FLEXORS				
Flexor carpi ulnaris	Ulna; medial epicondyle of humerus	Fifth metacarpal; pisiform and hamate	Flexes and adducts wrist; flexes forearm	Ulnar
Palmaris longus	Medial epicondyle of humerus	Palmar fascia	Tenses palmar fascia; flexes wrist	Median
Flexor carpi radialis	Medial epicondyle of humerus	First and second metacarpals	Flexes and abducts wrist	Median
Flexor digitorum profundus	Ulna	Distal phalanges 2–5	Flexes fingers and wrist	Median and ulnar
Flexor digitorum superficialis	Medial epicondyle of radius	Middle phalanges 2–5	Flexes fingers and wrist	Median
Flexor pollicis longus	Radius	Distal phalanx of thumb	Flexes thumb and wrist	Median
EXTENSORS				
Extensor carpi ulnaris	Ulna; lateral epicondyle of humerus	Metacarpal 5	Extends hand; adducts little finger	Radial
Extensor digitorum	Lateral epicondyle of humerus	Phalanges 2–5	Extends fingers and wrist	Radial
Extensor carpi radialis brevis	Lateral epicondyle of humerus	Metacarpal 3	Extends and abducts wrist	Radial
Extensor carpi radialis longus	Lateral supracondylar ridge of humerus	Metacarpal 2	Extends and abducts wrist	Radial
Extensor indicis	Ulna	Phalanx 2	Extends forefinger and wrist	Radial
Abductor pollicis longus	Posterior ulna and radius; interosseous membrane	Metacarpal 1	Abducts and extends thumb and wrist	Radial
Extensor pollicis longus	Dorsal surface of ulna	Base of thumb, second phalanx	Extends end of thumb	Radial
Extensor pollicis brevis	Dorsal surface of radius	Dorsal surface of thumb, first phalanx	Extends and abducts thumb; abducts wrist	Posterior interosseous

NOTE: These last two muscles cross the lateral surface of the wrist to form the **anatomical snuff box.** Extend the thumb laterally to see this structure. The radial artery passes through the snuff box; the pulse can be felt there.

TABLE 5.18 Muscles that move the chest wall: Breathing

NAME	ORIGIN	INSERTION	ACTION	INNERVATION
NOTE: These muscles are overlaid by the latissimus dorsi, trapezius, and the pectoralis, which are functionally part of the appendicular muscle division.				
External intercostals	Inferior border of rib	Superior border of rib	Draws adjacent ribs together	Intercostal
Internal intercostals	Inferior border of rib	Superior border of rib	Draws adjacent ribs together	Intercostal
Transversus thoracis	Lower one-third of sternum	Costal cartilage of true ribs (except first rib)	Depresses ribs in exhalation	Intercostal
Diaphragm	Xiphoid process, costal cartilages of lowest six ribs, lumbar vertebrae	Central tendon	Depresses floor of thoracic cavity in inhalation	Phrenic
Sternocleidomastoid	See MUSCLES THAT MOVE THE HEAD. If head acts as origin, then this muscle acts to elevate sternum and rib cage.			
Scalenes	Transverse processes of second to seventh cervical vertebrae	First two ribs	Elevates ribs in inhalation	C5–C8
Levatores costarum	Transverse processes of seventh cervical and first eleven thoracic vertebrae	Angle of rib immediately below origin	Elevates ribs in inhalation	Intercostal

TABLE 5.19 Muscles that move the abdominal wall

NAME	ORIGIN	INSERTION	ACTION	INNERVATION
External oblique	Lower eight ribs	Iliac crest, linea alba	Compresses abdominal contents	Intercostals 8–12, iliohypogastric, ilioinguinal
Internal oblique	Iliac crest, inguinal ligament, lumbodorsal fascia	Costal cartilages of last three or four ribs	Compresses abdominal contents	Same as external oblique
Transversus abdominis	Iliac crest, inguinal ligament, lumbar fascia, costal cartilages of last six ribs	Xiphoid process, linea alba, pubis	Compresses abdominal contents	Intercostals 7–12, iliohypogastric, ilioinguinal
Rectus abdominis	Pubic crest, symphysis pubis	Xiphoid process, costal cartilages of fifth, sixth, and seventh ribs	Flexes trunk, compresses abdominal contents	Intercostals 7–12

TABLE 5.20 Muscles of the pelvic floor: The pelvic diaphragm

NAME	ORIGIN	INSERTION	ACTION	INNERVATION
Levator ani	Posterior surface of pubis, ischial spine	Coccyx	Support pelvic organs. Supports pregnant uterus, participates in childbirth	Pudendal
Coccygeus (posterior continuation of levator ani)	Ischial spine	Coccyx, sacrum	Same as levator ani	Pudendal
Sphincter ani externus	Coccyx	Central tendon of perineum	Closes anal canal	Pudendal and S4
Sphincter urethrae	Pubic ramus	Central tendon of perineum	Constricts urethra	Pudendal
Ischiocavernosus	Ischial ramus	Corpus cavernosum	Compresses base of penis or clitoris	Pudendal
Transverse perinei	Ischial ramus	Central tendon of perineum	Supports pelvic floor	Pudendal
Bulbospongiosus (male)	Perineum and bulb of penis	Central tendon of perineum	Constricts urethra and erects penis	Pudendal
Bulbospongiosus (female)	Central tendon of perineum	Base of clitoris	Erects clitoris	Pudendal

TABLE 5.21 Muscles of the vertebral column: Muscles of erect posture

NAME	ORIGIN	INSERTION	ACTION	INNERVATION
NOTE: Muscles of the abdominal wall function as postural muscles.				
Iliopsoas	Postural muscle when femur acts as origin	See MUSCLES LOCATED IN THE ANTERIOR HIP		
ERECTOR SPINAE GROUP Composed of three muscle groups, each of which has subgroups. The three major groups are the laterally placed **iliocostalis,** the intermediately placed **longissimus,** and the medially placed **spinalis.**				
Iliocostalis **Lumborum** **Thoraois** **Cervicis**	Iliac crest and all ribs	Ribs or transverse processes roughly six vertebrae above origin	Extends trunk and neck, maintains erect posture, rotates trunk and neck	Dorsal rami of lumbar, thoracic, and cervical spinal nerves
Longissimus **Thoracis** **Cervicis** **Capitis**	Transverse processes of thoracic and lumbar vertebrae	Transverse processes roughly twelve vertebrae above origin, some ribs, and mastoid process of temporal bone	Extends trunk and neck, maintains erect posture, rotates trunk and head	Dorsal rami of lumbar, thoracic, and cervical spinal nerves
Spinalis **Thoracis** **Cervicis** **Capitis**	Spinous processes of upper lumbar and lower thoracic vertebrae	Spinous processes of upper thoracic vertebrae, cervical vertebrae and occipital bone	Extends trunk	Dorsal rami of lumbar and thoracic spinal nerves
Semispinalis	Transverse processes of seventh cervical and thoracic vertebrae	Spinous processes roughly six vertebrae above origin, occipital bone	Extends and rotates vertebral column and head	Dorsal rami of spinal nerves
Multifidus	Pelvic girdle, lumbar vertebrae, transverse processes of thoracic and lower cervical vertebrae	Spinous processes three vertebrae above origin	Extends and rotates trunk	Dorsal rami of lumbar, thoracic, and cervical spinal nerves
Quadratus lumborum	Posterior iliac crest and lower three lumbar vertebrae	Twelfth rib and transverse processes of top four lumbar vertebrae	Lateral flexion of trunk, pelvic extension	T12, L1

TABLE 5.22 Muscles located in the lateral hip

NAME	ORIGIN	INSERTION	ACTION	INNERVATION
Tensor fasciae latae	Anterior iliac crest	Through iliotibial band to lateral tibia	Tenses and abducts thigh	Superior gluteal

TABLE 5.23 Muscles located in the anterior hip

NAME	ORIGIN	INSERTION	ACTION	INNERVATION
Iliopsoas **Two components: Iliacus and psoas major**	Transverse processes of lumbar vertebrae, iliac fossa	Lesser trochanter of femur and iliopubic junction	Flexes and laterally rotates thigh, also flexes trunk	L1–L3
Rectus femoris	See MUSCLES LOCATED IN ANTERIOR THIGH			

TABLE 5.24 Muscles located in the posterior hip

NAME	ORIGIN	INSERTION	ACTION	INNERVATION
Gluteus maximus	Posterior iliac crest, sacrum, coccyx	Iliotibial tract and gluteal tuberosity of femur	Extends and rotates thigh laterally	Inferior gluteal
Gluteus medius	Lateral surface of ilium	Greater trochanter of femur	Abducts and rotates thigh medially	Superior gluteal
Gluteus minimus	Lateral surface of ilium	Greater trochanter of femur	Abducts and rotates thigh medially	Superior gluteal
Piriformis	Sacrum	Greater trochanter of femur	Abducts and rotates thigh laterally	S1–S2

TABLE 5.25 Muscles located in the anterior thigh

NAME	ORIGIN	INSERTION	ACTION	INNERVATION
QUADRICEPS FEMORIS GROUP				
Rectus femoris	Anterior inferior iliac spine	Tibial tuberosity via patellar tendon	Flexes thigh and extends leg	Femoral
Vastus lateralis	Greater trochanter and linea aspera	Same as rectus femoris	Extends leg	Femoral
Vastus medialis	Linea aspera of femur	Same as rectus femoris	Extends leg	Femoral
Vastus intermedius (located immediately posterior to rectus femoris)	Anterior surface of femur	Same as rectus femoris	Extends leg	Femoral

TABLE 5.26 Muscles located in the medial thigh

NAME	ORIGIN	INSERTION	ACTION	INNERVATION
ADDUCTOR GROUP				
Adductor brevis	Inferior pubic ramus	Linea aspera of femur	Adducts, rotates, and flexes thigh	Obturator
Adductor longus	Pubic crest and symphysis pubis	Linea aspera of femur	Adducts, rotates, and flexes thigh	Obturator
Adductor magnus	Ischial tuberosity, ischiopubic ramus	Linea aspera of femur	Adducts, rotates, and flexes thigh	Obturator
Gracilis	Symphysis pubis and pubic arch	Medial surface of tibia	Flexes leg and adducts thigh	Obturator
Pectineus	Pubic spine and iliopubic junction	Pectineal line of femur (distal to lesser trochanter)	Flexes and adducts thigh, rotates thigh laterally	Femoral

TABLE 5.27 Muscles located in the posterior thigh

NAME	ORIGIN	INSERTION	ACTION	INNERVATION
HAMSTRING GROUP				
Biceps femoris	Long head: Ischial tuberosity Short head: Linea aspera of femur	Lateral portion of head of fibula, lateral tibial condyle	Flexes leg and extends thigh	Tibial and peroneal
Semitendinosus	Ischial tuberosity	Proximal medial tibia	Flexes leg and extends thigh	Tibial
Semimembranosus	Ischial tuberosity	Medial condyle of tibia	Flexes leg and extends thigh	Tibial

TABLE 5.28 Muscles located in the lower leg

NAME	ORIGIN	INSERTION	ACTION	INNERVATION
Gastrocnemius	Lateral and medial tibial condyles, knee capsule	Calcaneus via Achilles tendon	Plantar flexes foot, flexes leg	Tibial
Soleus	Head of fibula, medial surface of tibia	Calcaneus via Achilles tendon	Plantar flexes foot	Tibial
Plantaris	Linea aspera of femur	Calcaneus via Achilles tendon	Plantar flexes foot, flexes leg	Tibial
Popliteus	Lateral condyle of femur	Posterior tibia	Flexes and medially rotates leg	Tibial
Peroneus brevis	Fibula	Metatarsal 5	Plantar flexes foot	Peroneal
Peroneus longus	Fibula and lateral condyle of tibia	Cuneiform 1; Metatarsal 1	Plantar flexes foot	Peroneal
Flexor hallucis longus	Shaft of fibula	Distal phalanx of great toe	Flexes great toe, plantar flexes foot	Tibial
Flexor digitorum longus	Posterior surface of tibia	Distal phalanges of four lateral toes	Flexes toes, plantar flexes foot	Tibial
Tibialis posterior	Interosseous membrane of tibia and fibula	Several tarsals and metatarsals	Plantar flexes foot	Tibial
NOTE: The tendons of the three preceding flexor muscles pass through the ankle just posterior and inferior to the medial malleolus. From posterior to anterior, the order of these tendons is ***T***. posterior, F. ***d***igitorum longus, and F. ***h***allucis longus, which has led to their being casually referred to as ***T***om, ***D***ick, and ***H***arry.				
Extensor hallucis longus	Shaft of fibula, interosseous membrane	Distal phalanx of great toe	Extends great toe, dorsiflexes foot	Deep peroneal
Extensor digitorum longus	Lateral tibial condyle, anterior fibular surface	Middle and distal phalanges of four lateral toes	Extends toes, dorsiflexes foot	Deep peroneal
Tibialis anterior	Lateral condyle and body of tibia	First metatarsal and first cuneiform	Dorsiflexes foot	Deep peroneal
Peroneus tertius	Fibula and interosseous membrane	Metatarsal 5	Dorsiflexes and everts foot	Deep peroneal

TABLE 5.29 The cranial nerves

NUMBER AND NAME	EXIT FROM SKULL	FUNCTION
I. Olfactory	Cribriform plate of ethmoid	Sensory: Olfaction. Rhythmic sensitivity follows hormonal cycles in females.
II. Optic	Optic foramen	Sensory: Vision. Probable efferents may regulate retinal metabolism and structural renewal.
III. Oculomotor	Orbital fissure	Motor: Rectus superior, rectus inferior, rectus medius, and obliquus inferior muscles. Sensory: Proprioception. Autonomic (parasympathetic): Muscles of iris, ciliary muscle to control lens.
IV. Trochlear	Orbital fissure	Motor: Obliquus superior muscle. Sensory: Proprioception.
V. Trigeminal		
Ophthalmic branch	Orbital fissure	Sensory: Cornea, upper eyelid, scalp, skin of upper face.
Maxillary branch	Foramen rotundum	Sensory: Palate and upper jaw, teeth and gums, nasopharynx, skin of cheek, lower eyelid, upper lip.
Mandibular branch	Foramen ovale	Sensory: Lower jaw, teeth and gums, anterior two-thirds of tongue, mucous membrane of cheek, skin of lower lip, chin, and ear. Motor: Muscles of chewing, throat, middle ear.
VI. Abducens	Orbital fissure	Motor: Rectus lateralis muscle. Sensory: Proprioception.
VII. Facial	Stylomastoid foramen and internal auditory meatus	Motor: Muscles of facial expression, throat middle ear. Sensory: Proprioception, taste (anterior two-thirds of tongue), palate. Autonomic (parasympathetic): Tear glands, salivary glands, and secretory glands in pharynx.
VIII. Vestibulocochlear	Internal auditory meatus	Sensory: Hearing (cochlear branch), balance (vestibular branch).
IX. Glossopharyngeal	Jugular foramen	Sensory: Posterior one-third of tongue, posterior pharynx, taste (posterior one-third of tongue), proprioception. Motor: Pharyngeal muscle. Autonomic (parasympathetic): Salivary glands, carotid sinus.
X. Vagus	Jugular foramen	Sensory: Inferior pharynx, larynx, internal organs. Motor: Posterior pharynx, larynx, tongue. Autonomic (parasympathetic): Thoracic and abdominal viscera.
XI. Accessory	Jugular foramen	Motor: Posterior pharynx, sternocleidomastoid, trapezius muscles. Sensory: Proprioception.
XII. Hypoglossal	Hypoglossal canal	Motor: Tongue and throat. Sensory: Proprioception.

TABLE 5.30 Spinal nerves and their branches

NERVE	SPINAL COMPONENT		INNERVATION
CERVICAL PLEXUS: C1, C2, C3, C4			
	Superficial cutaneous branches		
	Lesser occipital	C2, C3	Skin of scalp above and behind ear
	Greater auricular	C2, C3	Skin in front of, above, and below ear
	Transverse cervical	C2, C3	Skin of anterior aspect of neck
	Supraclavicular	C3, C4	Skin of upper portion of chest and shoulder
	Deep motor branches		
	Ansa cervicalis		
	Anterior root	C1, C2	Geniohyoid, thyrohyoid, and infrahyoid muscles of neck
	Posterior root	C3, C4	Omohyoid, sternohyoid, and sternothyroid muscles of neck
	Phrenic	C3–C5	Diaphragm
	Segmental branches	C1–C5	Deep muscles of neck (levator scapulae ventralis, trapezius, scalenus, and sternocleidomastoid)
BRACHIAL PLEXUS: C5, C6, C7, C8, T1			
	Axillary	Posterior cord (C5–C6)	Skin of shoulder; shoulder joint, deltoid and teres minor muscles
	Radial	Posterior cord (C5–C8, T1)	Skin of posterior lateral surface of arm, forearm, and hand; posterior muscles of brachium and antebrachium (triceps brachii, supinator, anconeus, brachioradialis, extensor carpi radialis brevis, extensor carpi radialis longus, extensor carpi ulnaris)
	Musculocutaneous	Lateral cord (C5–C7)	Skin of lateral surface of forearm; anterior muscles of brachium (coracobrachialis, biceps brachii, brachialis)
	Ulnar	Medial cord (C8, T1)	Skin of medial third of hand; flexor muscles of anterior forearm (flexor carpi ulnaris, flexor digitorum), medial palm and intrinsic flexor muscles of hand (profundus, third and fourth lumbricales)
T2, T3, T4, T5, T6, T7, T8, T9, T10, T11, T12:			
No plexus in these segments; branches run directly to intercostal muscles and skin of thorax.			
LUMBOSACRAL PLEXUS: L1, L2, L3, L4, L5, S1, S2, S3, S4, S5			
Lumbar	Iliohypogastric	T12–L1	Skin of lower abdomen and buttock; muscles of anterolateral abdominal wall (external abdominal oblique, internal abdominal oblique, transversus abdominis)
	Ilioinguinal	L1	Skin of upper median thigh, scrotum and root of penis in male and labia majora in female; muscles of anterolateral abdominal wall with iliohypogastric nerve
	Genitofemoral	L1, L2	Skin of middle anterior surface of thigh, scrotum in male and labia majora in female; cremaster muscle in male
	Lateral femoral cutaneous	L2, L3	Skin of anterior, lateral, and posterior aspects of thigh
	Femoral	L2–L4	Skin of anterior and medial aspect of thigh and medial aspect of lower extremity and foot; anterior muscles of thigh (iliacus, psoas major, pectineus, rectus femoris, sartorius) and extensor muscles of leg (rectus femoris, vastus lateralis, vastus medialis, vastus intermedius)
	Obturator	L2–L4	Skin of medial aspect of thigh; adductor muscles of lower extremity (external obturator, pectineus, adductor longus, adductor brevis, adductor magnus, gracilis)
	Saphenous	L2–L4	Skin of medial aspect of lower extremity
Sacral	Superior gluteal	L4, L5, S1	Abductor muscles of thigh (gluteus maximus, gluteus medius, tensor fasciae latae)
	Inferior gluteal	L5–S2	Extensor muscle of hip joint (gluteus maximus)
	Nerve to piriformis	S1, S2	Abductor and rotator of thigh (piriformis)
	Nerve to quadratus femoris	L4, L5, S1	Rotators of thigh (gemellus inferior, quadratus femoris)
	Nerve to obturator internus	L5–S2	Rotators of thigh (gemellus superior, internal obturator)
	Perforating cutaneous	S2, S3	Skin over lower medial surface of buttock
	Posterior femoral cutaneous	S1–S3	Skin over lower lateral surface of buttock, anal region, upper posterior surface of thigh, upper aspect of calf, scrotum in male and labia majora in female
	Sciatic	L4–S3	Composed of two nerves (tibial and common fibular); splits into two portions at popliteal fossa; branches from sciatic in thigh region to "hamstring muscles" (biceps femoris, semitendinosus, semimembranosus) and adductor magnus muscle
	Tibial (sural, medial, and lateral plantar)	L4–S3	Skin of posterior surface of leg and sole of foot; muscle innervation includes gastrocnemius, soleus, flexor digitorum longus, flexor hallucis longus, tibialis posterior, popliteus, and intrinsic muscles of the foot
	Common fibular (superficial and deep fibular)	L4–S2	Skin of anterior surface of the leg and dorsum of foot; muscle innervation includes peroneus tertius, peroneus brevis, peroneus longus, tibialis anterior, extensor hallucis longus, extensor digitorum longus, extensor digitorum brevis
	Pudendal	S2–S4	Skin of penis and scrotum in male and skin of clitoris, labia majora, labia minora, and lower vagina in female; muscles of perineum

TABLE 5.31 Events of the cardiac cycle

PHASE	ELECTRICAL EVENTS	MECHANICAL EVENTS	HEART SOUND
Late diastole		AV valves open; semilunar valves closed. Blood enters all chambers by passive filling from venae cavae and pulmonary veins.	
Atrial systole	P wave: SA node depolarizes, wave spreads throughout atria P-R interval: Wave of depolarization reaches SA node. Typical P-R interval (beginning of P-wave to onset of next deviation from baseline) is <0.2 seconds	Atria contracted in response to depolarizing signal. Blood engorges ventricles, adding to stretch of ventricular walls.	
Isometric ventricular contraction	QRS complex: Depolarization of SA node. Bundle of His, and Purkinje fibers spread depolarization through valve ring into ventricular muscle	Ventricles contract in response to depolarizing signal. Papillary muscles relax, allowing AV valves to close. Typically, mitral closure slightly precedes tricuspid closure. Reverberation of blood against valve cusps produces low pitched "lub" of first heart sound. With all valves closed, ventricular pressure rises.	First heart sound (may be split with mitral component preceding tricuspid component).
Ventricular ejection	S-T segment: Entire ventricle is uniformly depolarized	Ejection begins when ventricular pressures exceed back pressures in aorta and pulmonary trunk. Semilunar valves open, blood from this cycle enters aorta and pulmonary trunk. Maintained depolarization during S-T segment permits efficient, coordinated ventricular emptying.	
Isometric ventricular relaxation	T-wave	Repolarization wave spreads through ventricles, permitting relaxation. As ventricular pressures drop below those of aorta and pulmonary trunk, semilunar valves close. Typically, aortic semilunar closes slightly before pulmonary semilunar. Reverberation of blood against closed valve cusps creates higher pitched "dub" of second heart sound. Lowered intraventricular pressures permit papillary muscles to pull AV valves open. Ventricular filling begins.	Second heart sound (typically split with pulmonary component slightly following aortic component, especially during inhalation).

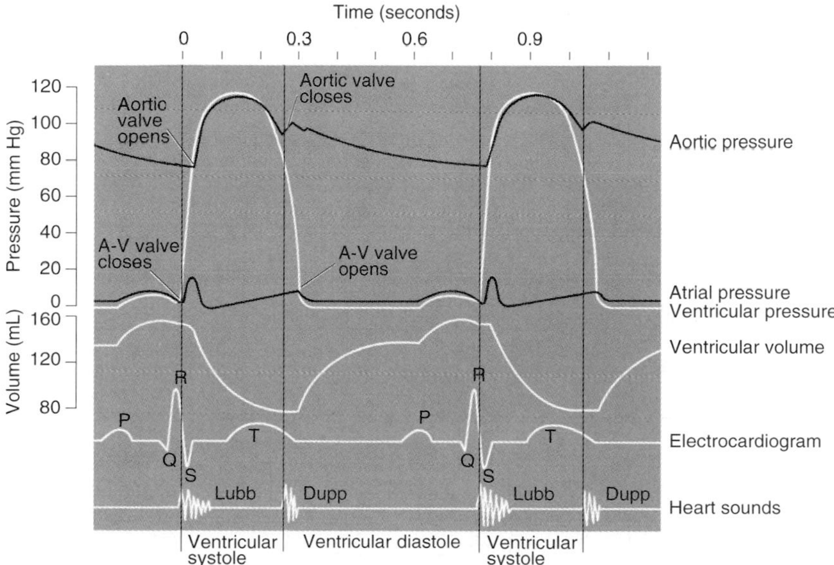

Figure 5-3
A Graph of Changes That Occur in Left Ventricle During a Cardiac Cycle

TABLE 5.32 Major blood vessels and their branches

MAJOR ARTERY	MAIN BRANCHES
Ascending aorta	Coronary arteries (right and left)
Aortic arch	Innominate (brachiocephalic) Left subclavian Left common carotid
Brachiocephalic innominate	Right subclavian Right common carotid
Common carotid (right and left)	Internal carotid External carotid
Subclavian (right and left)	Vertebral (right and left) Axillary (continuation of subclavian)
Axillary	Brachial (continuation of axillary)
Brachial	Radial Ulnar
Radial and ulnar	Palmar arches (superficial and deep)
Cerebral arterial circle (Circle of Willis)	Vertebrals join in cranium to form basilar artery, which then divides to form left and right posterior cerebral arteries. Internal carotids, upon entering cranium, become left and right anterior cerebral arteries. A pair of posterior communicating arteries and an anterior communicating artery join the cerebrals to form an arterial anastomosis, the circle of Willis.
Descending aorta	Intercostal arteries and spinal branches Celiac trunk (branches to hepatic, splenic, and right and left gastric arteries) Mesenteric (superior and inferior) Renal (right and left) Gonadal (spermatic or ovarian; right and left) Parietal branches to diaphragm, dorsal skin and skeletal muscles, and spinal cord Common iliac (right and left)
Common iliac	Internal iliac (or hypogastric; right and left) External iliac (right and left)
External iliac	Femoral (right and left)
Femoral	Popliteal (right and left)
Popliteal	Tibial (anterior and posterior; right and left)
Tibial	Plantar arches

MAJOR VEIN	COMMENT
UPPER EXTREMITY (RIGHT AND LEFT)	
Palmar arch (superficial and deep)	
Medial cubital	Connects cephalic and basilic
Median antebrachial	Median antebrachial and median cubital flow into basilic
Radial and ulnar	Radial and ulnar flow into brachial
Basilic and brachial	Basilic and brachial flow into axillary
Cephalic	Cephalic and axillary flow into subclavian
Axillary (continuation of brachial)	Cephalic and axillary flow into subclavian
Subclavian (continuation of axillary)	Flows into innominate (brachiocephalic)

TABLE 5.32 Major blood vessels and their branches (continued)

MAJOR VEIN	COMMENT
LOWER EXTREMITY (RIGHT AND LEFT)	
Plantar arch	
Dorsal venous arch	
Anterior tibial	Anterior and posterior tibials unite to form popliteal
Posterior tibial	Anterior and posterior tibials unite to form popliteal
Small saphenous	Flows into popliteal
Popliteal	Popliteal and peroneal unite to form femoral
Peroneal	Popliteal and peroneal unite to form femoral
Femoral	Femoral and great saphenous unite to form external iliac
Great saphenous	Femoral and great saphenous unite to form external iliac
External iliac	External and internal iliacs unite to form common iliac
Internal iliac	External and internal iliacs unite to form common iliac
Common iliac	Flows into inferior vena cava
ABDOMEN	
Lumbar (several pairs)	Flows into inferior vena cava and azygous system
Gonadal (spermatic or ovarian; right and left)	Flows directly into inferior vena cava
Renal (right and left)	Flows directly into inferior vena cava
Suprarenal (right and left)	Flows directly into inferior vena cava
Hepatic	Flows directly into inferior vena cava
Mesenteric (superior and inferior)	Flows into hepatic portal system
Splenic	Flows into hepatic portal system
Gastroepiploic (right and left)	Flows into hepatic portal system
Hepatic portal	Conveys blood to liver; hepatic vein flows from liver
THORAX	
Left intercostal	Flows into hemiazygos
Hemiazygos	Flows into azygos
Accessory hemiazygos	Flows into azygos
Right intercostal	Flows into azygos
Azygos	Flows into inferior vena cava
Coronary (right and left)	Flows into right atrium of heart
HEAD AND NECK	
Superior sagittal sinus	
Inferior sagittal sinus	Flows into straight sinus
Straight sinus	Flows into transverse sinus
Cavernous	Flows into petrosal sinus

TABLE 5.32 Major blood vessels and their branches (continued)

MAJOR VEIN	COMMENT
Petrosal sinus (right and left)	Flows into transverse sinuses
Transverse sinuses (right and left)	Flows into sigmoid sinuses
Sigmoid sinuses (right and left)	Flows into internal jugular vein
Internal jugular	
External jugular	
Vertebral (right and left)	
Brachiocephalic innominate (right and left)	Flows into superior vena cava
Superior vena cava	Flows into right atrium of heart

FETAL SYSTEM

Placenta ➛ Umbilical vein ➛ Ductus venosus (bypasses liver) ➛ Inferior vena cava ➛ Right atrium of fetus ➛ Mostly through foramen ovale ➛ Left atrium ➛ Left ventricle ➛ Mostly to fetal head ➛ Return to right atrium ➛ Mostly to right ventricle ➛ Pulmonary trunk ➛ Mostly through ductus arteriosus ➛ Descending aorta ➛ Common iliac arteries ➛ Internal iliac arteries ➛ Umbilical arteries ➛ Placenta

TABLE 5.33 Major hormones of the pituitary gland

HORMONE	CHEMICAL STRUCTURE	TARGET	REGULATION	MAJOR ACTION
ANTERIOR PITUITARY (ADENOHYPOPHYSIS)				
Growth Hormone (GH, Somatotropin)	Protein	General	GH Releasing Hormone from hypothalamus	Enhances protein anabolism, fat catabolism; enhances growth, wound healing, positive nitrogen balance
Prolactin (Prl)	Protein	Breast tissue	Inhibited by dopamine (a prolactin-inhibiting hormone) from hypothalamus	In female, mimics many actions of GH during pregnancy; enhances breast tissue anabolism for lactation
Adrenocorticotropic hormone (ACTH)	Polypeptide	Adrenal cortex	Corticotrophin releasing hormone from hypothalamus	Promotes secretion of glucocorticosteroids by adrenal cortex
Endorphins (several)	Peptide	Central nervous system neurons	Neural activity in hypothalamus in response to stress and probably suckling	Inhibits transmission of pain impulses; enhances feeling of well-being
Thyroid stimulating hormone (TSH)	Glycoprotein	Thyroid gland	Thyroid releasing hormone (TRH) from hypothalamus	Stimulates release of thyroid hormones
Follicle stimulating hormone (FSH)	Glycoprotein	Gonads	Gonadotropin releasing hormone (GnRH) from hypothalamus	Female: Maturation of ovarian follicle; estrogen secretion Male: Sperm production
Luteinizing hormone (LH)	Glycoprotein	Gonads	Gonadotropin releasing hormone (GnRH) from hypothalamus	Female: Rupture of follicle; ovulation Male: Testosterone secretion
POSTERIOR PITUITARY (NEUROHYPOPHYSIS)				
Antidiuretic hormone (ADH, Vasopressin)	Peptide	Kidney tubules	Neural activity in hypothalamus in response to brain osmoreceptors; stress	Increase water retention; elevation of blood pressure
Oxytocin	Peptide	Breast tissue, uterus	Neural activity in hypothalamus in response to suckling, uterine stimulation	Let down of milk in lactating breast; uterine smooth muscle contractions

NOTE: The above-named list of hormones is not an exhaustive list of substances now known to be secreted by the pituitary gland. In addition, the listed hormones are known to have several actions, many of which are also not included.

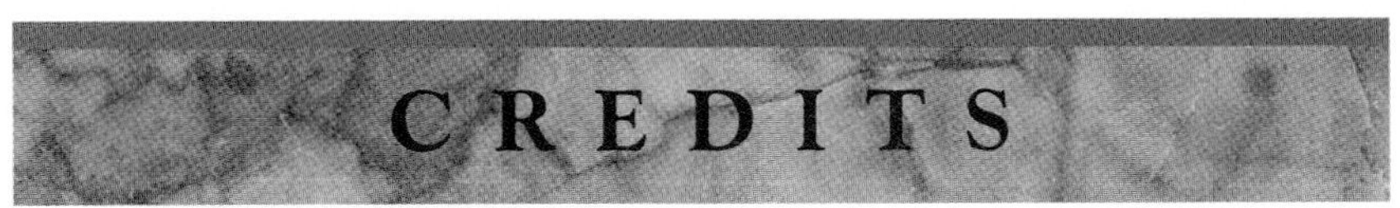

CREDITS

Chapter 1

Chapter Opener 1: © Ed Reschke; 1.1, 1.2, 1.3, 1.4, 1.5, 1.6, 1.7, 1.8, 1.9, 1.10, 1.11, 1.12, 1.13, 1.14, 1.15, 1.16, 1.17, 1.18, 1.19, 1.20, 1.21, 1.22, 1.23, 1.24, 1.25, 1.26, 1.27, 1.28, 1.29, 1.30, 1.31, 1.32, 1.33, 1.34, 1.35, 1.36, 1.37, 1.38, 1.39, 1.40, 1.41, 1.42, 1.43, 1.44, 1.45, 1.46, 1.47, 1.48, 1.49, 1.50, 1.51, 1.52: © Ed Reschke; 1.53: Courtesy Bruce Wingard; 1.54, 1.55, 1.56, 1.57A-B, 1.58, 1.59, 1.60, 1.61, 1.62, 1.63, 1.64, 1.65, 1.66, 1.67, 1.68A: © Ed Reschke; 1.68B: © Camera M.D. Studios; 1.69, 1.70, 1.71, 1.72, 1.73A-B, 1.74, 1.75, 1.76, 1.77, 1.78, 1.79, 1.80A: © Ed Reschke; 1.80B: © Visuals Unlimited, Inc.; 1.81, 1.82, 1.83, 1.84A-C, 1.85, 1.86, 1.87, 1.88, 1.91, 1.92, 1.93, 1.94, 1.95, 1.96, 1.97, 1.98, 1.99, 1.101, 1.102, 1.103, 1.104, 1.105, 1.106: © Ed Reschke

Chapter 2

Chapter Opener 2: © SPL/Photo Researchers, Inc.; 2.1, 2.2, 2.3, 2.4, 2.5, 2.6, 2.7, 2.8, 2.9, 2.10, 2.11, 2.12, 2.13, 2.14, 2.15A-B, 2.16, 2.17, 2.18, 2.19A-B, 2.21, 2.22, 2.23, 2.24, 2.25, 2.26, 2.27; Douglas Eder, Shari Lewis Kaminsky and John Bertram; 2.28: © Dennis Strete; 2.29, 2.30: Douglas Eder, Shari Lewis Kaminsky, and John Bertram; 2.31: © Dennis Strete; 2.32, 2.33, 2.34, 2.35, 2.36, 2.37, 2.38: Douglas Eder, Shari Lewis Kaminsky, and John Bertram

Chapter 3

Chapter Opener 3: © SPL/Photo Researchers, Inc.

Chapter 4

Chapter Opener 4: © SPL/Photo Researchers, Inc.; 4.1, 4.2, 4.3, 4.4, 4.5, 4.6, 4.7, 4.8, 4.9, 4.10, 4.11, 4.12, 4.13A-B, 4.14, 4.15, 4.16, 4.17, 4.18, 4.19, 4.20, 4.21, 4.22, 4.23, 4.24, 4.25A-B, 4.26A-B, 4.27, 4.28A-B, 4.29, 4.30A-B, 4.31, 4.32, 4.33A-B, 4.34, 4.35, 4.36, 4.37A-B, 4.38, 4.39, 4.40, 4.41, 4.42, 4.43, 4.44, 4.45, 4.46: Douglas Eder, Shari Lewis Kaminsky, and John Bertram

Chapter 5

Chapter Opener 5: Astrid & Hans-Frieder Michler/SPL/Photo Researchers, Inc.

INDEX